I0049528

Food Packaging Based on Nanomaterials

Food Packaging Based on Nanomaterials

Special Issue Editors

Amparo López-Rubio
Maria Jose Fabra
Marta Martínez-Sanz

MDPI • Basel • Beijing • Wuhan • Barcelona • Belgrade

MDPI

Special Issue Editor
Amparo López-Rubio, Maria Jose Fabra and Marta Martínez-Sanz
CSIC—Instituto de Agroquimica y Tecnología de los Alimentos (IATA)
Spain

Editorial Office
MDPI
St. Alban-Anlage 66
4052 Basel, Switzerland

This is a reprint of articles from the Special Issue published online in the open access journal *Nanomaterials* (ISSN 2079-4991) in 2018 (available at: https://www.mdpi.com/journal/nanomaterials/special_issues/nano_food_packaging)

For citation purposes, cite each article independently as indicated on the article page online and as indicated below:

LastName, A.A.; LastName, B.B.; LastName, C.C. Article Title. *Journal Name* **Year**, *Article Number*, Page Range.

ISBN 978-3-03897-501-4 (Pbk)
ISBN 978-3-03897-502-1 (PDF)

Cover image courtesy of Amparo López-Rubio.

© 2018 by the authors. Articles in this book are Open Access and distributed under the Creative Commons Attribution (CC BY) license, which allows users to download, copy and build upon published articles, as long as the author and publisher are properly credited, which ensures maximum dissemination and a wider impact of our publications.

The book as a whole is distributed by MDPI under the terms and conditions of the Creative Commons license CC BY-NC-ND.

Contents

About the Special Issue Editors

Amparo López-Rubio (Dr.) has been a tenured researcher at IATA-CSIC since 2014. She received her Ph.D. in Food Science (2006) at the Polytechnic University of Valencia (UPV, Spain) with a thesis in food packaging, awarded with the "Premio Extraordinario de Tesis Doctoral" from UPV and with the second prize to best doctoral thesis by the "Specialized Polymer Group" from the Spanish Royal Society of Physics (SRSP). After pre- and postdoctoral training in top research institutions around the world (KTH, Sweden; ANSTO, Australia; Hasylab, Germany; ANL, USA) where she broadened her knowledge about advanced materials characterization tools, she returned to IATA to start a research line on nanoencapsulation of bioactive compounds. Her research is focused on understanding the relationship between structure and functionality of food components and materials for a rational design of their combination. She has published over 140 scientific articles in peer-reviewed journals (H-index of 32 on Web of Science), eight patents (three of them under exploitation), as well as numerous book chapters and conference proceedings. She is the editor of several books and Special Issues and has been appointed Chair of the FWO Expert Panel W&T6 (Belgium). She is Fellow of the SRSP and member of the Nanotechnology group from AECOSAN. She has established numerous national and international collaborations and has secured public and private funding worth over €900.000 as a PI or co-PI while working at IATA.

Maria Jose Fabra (Dr.) is a postdoctoral researcher at the Food safety and Preservation Department from the Institute of Agrochemistry and Food Technology (IATA-CSIC, Spain). Throughout her scientific career, she has developed a multidisciplinary cutting-edge background, which spans from the characterization of foods to the development of highly functional biopolymers by means of nanotechnology using a broad range of analytical techniques. Her current research interests focus on the development of enzymatically active biopolymers for food applications. Her research activities have resulted in more than 80 SCI publications in high-impact journals, 18 book chapters, and more than 80 communications in scientific congresses. She is co-editor of two books based on biopolymers and nanotechnology. She is actively collaborating with a number of national and international research groups.

Marta Martínez-Sanz (Dr.) is a Chemical Engineer (UPV) and holds a Ph.D. in Food Science (UPV). She carried out her PhD at the Institute of Agrochemistry and Food Technology (IATA-CSIC), specializing on the synthesis and characterization of bio-based nanofillers extracted from renewable resources, such as plant-derived and bacterial cellulose nanocrystals, as well as on the development of novel routes to incorporate them into polymeric matrices. After that, she worked for three years as a joint postdoctoral research fellow in the Australian Nuclear Science and Technology Organisation (ANSTO) and the University of Queensland (UQ), where her research focused on the investigation of the multi-scale structure of plant cell walls and model systems utilizing small-angle scattering techniques and diffraction methods in combination with complementary microscopy, spectroscopy, and rheology methods. Additionally, part of her research was related to the structural characterization of food-based systems. In 2017, she was a granted a Juan de la Cierva postdoctoral fellowship and returned to Spain to join the Packaging Group at the IATA-CSIC, where she started working on the production and characterization of polysaccharides extracted from renewable resources such as aquatic resources (seaweed, algae, and plants), as well as from food and agriculture-derived wastes. Overall, her research career is focused on the multi-scale structural investigation of bio-based materials and food-based systems, making use of a wide range of techniques and advanced characterization tools. Her scientific activity has led to more than 40 publications in scientific journals and book chapters, five patents, and participation in more than 15 international conferences of relevance for different research areas.

Preface to "Food Packaging Based on Nanomaterials"

The use of nanotechnologies in the food-packaging area has opened up a number of possibilities derived from the inherent characteristics of nanoadditives, which can either improve relevant properties of neat polymers (such as barrier or mechanical properties) or introduce new functionalities (for active and bioactive packaging applications or even for sensing). This is an exciting and rapidly growing field of study, and very interesting developments are unfolding. Although the aim of these novel materials is to improve packaged food quality and safety, the toxicological effects derived from their potential migration from the polymer structures is also under consideration.

Amparo López-Rubio, Maria Jose Fabra, Marta Martínez-Sanz
Special Issue Editors

nanomaterials

MDPI

Review

Recent Developments in Food Packaging Based on Nanomaterials

Yukun Huang [1], Lei Mei [2], Xianggui Chen [1] and Qin Wang [1,2,*]

1   School of Food and Bioengineering, Xihua University, Chengdu, Sichuan 610039, China;
    huangyukun@mail.xhu.edu.cn (Y.H.); chen_xianggui@mail.xhu.edu.cn (X.C.)
2   Department of Nutrition and Food Science, College of Agriculture and Natural Resources,
    University of Maryland, College Park, MD 20740, USA; leimei@umd.edu
*   Correspondence: wangqin@umd.edu; Tel.: +1-301-405-8421

Received: 31 August 2018; Accepted: 8 October 2018; Published: 13 October 2018

Abstract: The increasing demand for high food quality and safety, and concerns of environment sustainable development have been encouraging researchers in the food industry to exploit the robust and green biodegradable nanocomposites, which provide new opportunities and challenges for the development of nanomaterials in the food industry. This review paper aims at summarizing the recent three years of research findings on the new development of nanomaterials for food packaging. Two categories of nanomaterials (i.e., inorganic and organic) are included. The synthetic methods, physical and chemical properties, biological activity, and applications in food systems and safety assessments of each nanomaterial are presented. This review also highlights the possible mechanisms of antimicrobial activity against bacteria of certain active nanomaterials and their health concerns. It concludes with an outlook of the nanomaterials functionalized in food packaging.

Keywords: nanomaterials; food packaging; inorganic nanoparticles; organic biopolymer composites; synthesis; activity; application; safety assessment; mechanisms

---

1. Introduction

Nanoscience and nanotechnology have become exciting fields of research and development since its introduction by Richard Feynman in 1959 [1]. At the heart of research in these fields are the synthesis, characterization, modeling and applications of new materials with nanometer-scale dimensions, at least one of the three external dimensions ranging from approximately 1 nm to 100 nm, which are called "nanomaterials". There are numerous nanomaterials that have been reported in many prior studies, generally divided into the so-called zero-dimensional (e.g., nanoparticles (NPs): quantum dots, nanoclusters and fullerenes), one-dimensional (e.g., nanotubes and nanorods), two-dimensional (e.g., thin films), and three-dimensional (e.g., nanocomposites and nanofibers) nanomaterials [2]. These materials have exhibited unusual mesoscopic properties, including high surface area, fine particle size, high reactivity, high strength and ductility, which are the reasons that nanomaterials are frequently applied in a diversified range of industrial fields [3,4]. As the researches of multi-disciplinary areas move along, nanomaterials are advancing with wide applications to electronic, optical and magnetic devices, biology, medicine, energy, defense and so on. In addition, their developments in food and agriculture industries are nearly similar to their modernization in medicine delivery and pharmaceutical areas [5,6].

In recent years, owing to the unique properties of nanomaterials other than their bulk counterparts mainly covering physical, chemical and biological properties, studies on the synthesis, characterization, applications and assessments of these materials have promoted the scientific advancement to grow and alter the entire agrifood area [7,8]. Specifically, many reports have focused on the potential applications of nanomaterials as participants to assure food quality, improve packaging and produce food products with altered function and nutrition [1,4,5,9,10].

Packaging is a key component of each stage in the food industry; however, its permeable nature is the major defect in conventional food packaging materials. There are no packaging materials fully resisting water-vapors and atmospheric gases [2]. Moreover, participants along with the food supply chain seek novel, cost-effective, eco-friendly and resourceful food packaging systems to protect and monitor the quality of packaged foods, which is made possible with committed food safety, quality and traceability. As a result, there are several critical factors driving the innovation of food packaging materials to be continuously excavated. On the one hand, food packaging facilitates storage, handling, transport and protection of food from environmental pollution and other influences, and meets the increasing demands of the market, especially related to consumer preference for nutritious and high-quality food products [11]. Some bionanocomposites materials are designed to improve the functional characteristics of general food packaging, such as barrier performance, mechanical strength and thermal stability, and other nanomaterials can incorporate bacteriostatic agents, antioxidants, plant extracts and enzymes to lengthen shelf-life of food products [12]. On the other hand, to date, the majority of materials used in packaging industries are non-biodegradable petroleum-based plastic polymer materials (approximately 8% of the global gas production and fossil feedstock is used to yield synthetic polymers), which in turn, denote a serious problem on the universal environment [13]. The advancement of renewable or green packaging has potentials to reduce the negative environmental impacts caused by the synthetic packaging by using biodegradable or edible materials, plant extracts, and nanocomposite materials [11]. The following two types of materials [14–19] are in focus: (1) inorganic and metal nanoparticles (nano-clay, montmorillonite nanoparticles, halloysite nanotubes, AgNPs, ZnO-NPs and CuO-NPs, et al.); (2) plant extracts (milk thistle extract, green tea extract, etc.) mixtures incorporated in biopolymers (chitosan, cellulose, starch, etc.).

Furthermore, the enormous potential of nanotechnology has received attention from researchers in multi-disciplinary areas to develop promising and desirable materials in food packaging systems. On the whole, the applications of nanocomposite materials for food packaging reported in the recent three years are divided into three main functions, i.e., improved, smart and active food packaging [2]. Firstly, improved packaging is that the utilization of nanoparticles in the bionanocomposite materials improves their mechanical and barrier properties, including elasticity, gas barrier characteristics (barrier against oxygen, carbon dioxide, and flavor compounds diffusion) and stability under different temperature and moisture conditions [12]. Secondly, smart (intelligent) packaging performs in terms of information feedback and marketing on real-time quality of packaged food products and also performs as a guard against fraud and fake products and an indicator of the situation of exposure to certain adverse factors such as insufficient temperatures or high oxygen levels [20,21]. Thirdly, active packaging offers protection and preservation grounded on mechanisms activated by inherent and/or acquired factors (antimicrobial activity, biodegradable activity), and achieves the reduction in loss of food products due to extension of their shelf-life [22]. Though there have been considerable studies on novel nanomaterials applications in food packaging reported every day, most materials are still in the stage of feasibility and demonstration studies, and employments in food packaging field are yet to receive approval concerning their safety issues, which could be caused by the migrations of nanomaterials from packaging to food matrix [23]. Moreover, the absorption, distribution, metabolism and excretion as well as toxicological assessment of nanoparticles in food intake of humans are important research focuses [24]. Thus, as can been seen, the use of nanomaterials in the food industry opens up multiple possibilities originating from the inherent features of nano-additives, which are either an improvement of the original polymer properties (e.g., barrier or mechanical properties) or introduction of new functionalities (e.g., active and bioactive packaging or sensing and monitoring). This is an emerging and evolutionary area involving multidisciplinary studies.

This review references more than 170 articles published in the recent three years and summaries the up-to-date developments of nanomaterials applied in the food packaging field, presenting a comprehensive review of various nanostructures and related technologies used to construct functional food packaging systems. The contents of this article mainly concentrate on synthesis methods, physical and chemical properties and biological activity, applications in food systems and safety assessments of different types of nanomaterials. This review also highlights the possible mechanism of some characteristics, such as antimicrobial activity against bacteria and improved reduction and stabilization properties of certain active nanomaterials. In the last part, an outlook of the nanomaterials functionalized in food packaging is included.

2. Inorganic and Metal Oxide Nanomaterials Applied in Food Packaging

Generally, nanomaterials applied in food packaging can be classified into two categories: inorganic and organic materials. For the former materials, metals and metal oxides and clay nanoparticles incorporated into bionanocomposite films and nanofibers can be considered [25,26]. Besides common bacteriostatic silver nanoparticles, some of the inorganic agents, like oxidized nanoparticles including CuO, ZnO, TiO_2, MgO and Fe_3O_4, have attracted great interest due to their resistance to the rough processing conditions and enhancement of strong inhibition against foodborne pathogens. As for the other materials like various clays, they could offer resistance to gases and water vapor, and improve the mechanical strength of biopolymers [2,27]. The second group is organic materials including, but not limited to, phenols, halogenated compounds, quaternary ammonium salts, plastic polymers, plus natural polysaccharide or protein materials such as chitosan, chitin, zein and whey protein isolates, which have lately been highly regarded [28,29].

2.1. Silver-Based Nanoparticles

So far, in all kinds of nanoparticles developed and characterized, silver-based nanoparticles (NPs) have taken an important place due to their inherent feature of antimicrobial activity even in solid-state samples, and have therefore been used as bacteriostatic agents from ancient times. Silver-salts materials also have an inhibition effect on the growth of diverse pathogens affecting human health, such as those in films, catheters, burns, cuts and wounds to protect them from infection [7]. Silver-based particles in nanoscale include silver nanoparticles (AgNPs), silver nanocluster (AgNC) and silver-based alloy materials [30,31].

2.1.1. Synthesis Methods

AgNPs is one of the most studied and applied antimicrobial agents because of its broad-spectrum antimicrobial activity against microorganisms. The traditional solvothermal synthesis methods of AgNPs-functionalized packaging materials which usually require physical and chemical preparations of synthesizing and immobilizing, however, seem to be very expensive and hazardous and not environmentally friendly. This method has been gradually discarded for the tedious and complicated procedure. Interestingly, AgNPs prepared through biological synthesis exhibit high solubility, yield and stability. Additionally, it is simpler, faster, more environmentally friendly and dependable, and is recognized as a green approach to produce AgNPs with well-defined morphology and size under optimal conditions in favor of application in food packaging [32]. Chu et al. prepared antimicrobial active poly (lactic acid) (PLA)-based films with alloy of AgNPs and zinc oxide nanoparticles (ZnO-NPs) through a solvent volatilizing method [33]. Tao et al. developed a convenient and efficient biosynthesis method to synthesize AgNPs-silk/poly (vinyl alcohol) (PVA) bionanocomposite film by blending AgNPs with PVA [34]. Shao et al. reported a new green chemistry synthetic method of sodium alginate-AgNPs composite by using sodium alginate as a stabilizing agent and ascorbic acid as a reducing agent [35]. Narayanan and Han presented an immobilization method of borate-stabilized AgNPs as nanofillers in dual-crosslinked polymers comprised of PVA and sodium alginate at different ratios [36]. Patra et al. produced a phyto-mediated biosynthesis of AgNPs through utilizing the water extract of watermelon rind under light exposure at room temperature, obtaining prepared AgNPs with an average size of 110 nm and surface plasmon resonance of 425 nm [37]. Azlin-Hasim et al. studied the capacity of a layer-by-layer strategy to prepare low-density polyethylene (LDPE) active films with silver nanoparticles coated for food packaging applications [38]. It is found that the green chemistry synthesis for the silver-based nanoparticles is highly effective and displays high potentials.

2.1.2. Physical, Chemical Properties and Biological Activity

The physical and chemical properties of nanoparticles are important for their action, efficacy, bio-distribution and safety. Accordingly, characterizations of nanomaterial are crucial to evaluate functions of the developed particles [32]. Characterizations are performed using a group of analytical techniques, including transmission electron microscopy (TEM), scanning electron microscopy (SEM), UV-Vis spectroscopy, Fourier transform infrared spectroscopy (FTIR), X-ray diffractometry (XRD), X-ray photoelectron spectroscopy (XPS), dynamic light scattering (DLS), atomic force microscopy (AFM), thermogravimetric analysis (TGA) and differential scanning calorimetric (DSC), to investigate their physical and chemical properties. Those properties include size and size distribution, surface chemistry, particle morphology, coating/capping, particle composition, agglomeration, dissolution rate, thermo-mechanical behavior, rheological property and particle reactivity in solution. It is equally important that the biological activities of nanomaterials are to be examined for ensuring their claimed antimicrobial property and safety concerns. Tao et al. found that PVA film coated by AgNPs-silk showed superior stability, mechanical performance and good antimicrobial activity inhibiting both Gram-positive and Gram-negative bacteria [34]. Arfat et al. developed the bionanocomposite films based on fish skin gelatin and bimetallic Ag-Cu nanoparticles (Ag-Cu NPs). The films were characterized to have improved mechanical property and low transparency, thermal stability, yellowness and high antibacterial activity against both Gram-positive and Gram-negative bacteria [39]. Jafari et al. studied the effect of chitin nanofiber on the morphological and physical properties of chitosan/silver nanoparticle bionanocomposite films, and concluded that AgNPs had dramatically improved the barrier and mechanical properties, but showed a negative impact on color properties [40]. Ramachandraiah et al. demonstrated a higher antioxidant activity of the biosynthesized AgNPs from persimmon byproducts and incorporation in sodium alginate thin films [41].

2.1.3. Applications in Food Systems

Because of the aforementioned unique properties, AgNPs have been widely used in the health care industry, house-hold utensils, food storage, environmental and biomedical applications. Herein, it is interesting to emphasize the applications of AgNPs in food systems, including antibacterial, antifungal, antioxidant, anti-inflammatory, antiviral, anti-angiogenic and anti-cancer. Heli et al. reported that the exposure of corrosive vapor (ammonia) remarkably reduced the population density of AgNPs embedded into bacterial cellulose, causing a large distance between the residual nanoparticles and a decrease in the UV-Vis absorbance related to the plasmonic properties of AgNPs [42]. This material exhibited color changes from amber to light amber upon corrosive vapor exposure, and from amber to a grey or taupe color upon fish or meat spoilage exposure, which opened up an innovative approach and capability in gas sensing to act as a smart packaging for monitoring fish or meat spoilage exposure. Tavakoli et al. investigated the effect of nano-silver packaging in increasing the shelf-life of nuts in an in vitro model, showing an important effect on extending the shelf-life of nuts with the highest shelf-life of hazelnuts, almonds, pistachios and walnuts extended to 18, 19, 20 and 18 months, respectively [43]. Deus et al. evaluated the effect of an edible film coated with nano-silver on the quality of turkey meat during modified atmosphere and vacuum-sealed packaging for 12 days of storage [44]. Ahmed et al. created PLA composite films by loading bimetallic silver-copper nanoparticles and cinnamon essential oil into polymer matrix through compression molding technique, which was utilized in the chicken meat packaging, revealing a new direction of active food packaging to control the pathogenic and spoilage bacteria related to fresh chicken meat [45].

2.1.4. Safety Assessments

On account of gaps in understanding toxicology of nanomaterials, the development of their applications is related to safety concerns. In case of food contact bio-nanocomposite materials, the first steps of consumers' exposures are the migrations of nanoparticles from packaging to food products. Thus, in order to estimate the risk, we need to know the possibilities of nanoparticles released from food contact materials [46]. Gallocchio et al. evaluated silver migration from a commercially available food packaging containing AgNPs into chicken meatballs under plausible domestic storage conditions, and tested the contribution of this packaging to restrict food spoilage bacteria proliferation [47]. The results showed that the migration was slow and no significant difference in the analyzed bacteria levels between meatballs stored in AgNPs plastic bags and control bags. Tiimob et al. tested the release of eggshell-silver tailored copolyester polymer blend film exposed to water and food samples by atomic absorption spectroscopy (AAS) analysis, showing that AgNPs was not released in chicken breast or distilled water until 168 and 72 h, respectively [48]. Su et al. estimated the effects of organic additives (Irganox 1076, Irgafos 168, Chimassorb 944, Tinuvin 622, UV-531 and UV-P) on the release of silver from nanosilver-polyethylene composite films to an acidic food simulant (3% acetic acid) by detection using inductively coupled plasma mass spectrometry (ICP-MS) and found that additives influenced silver release through two synchronous processes: (1) reactions between silver and organic additives promoted release of silver from the composite film to an acidic food simulant; (2) inhibition or promotion of silver release was influenced by silver oxidation [49]. High humidity and temperature treatment of the prepared films were suggested to markedly enhance silver release by promoting oxidation. Hosseini et al. measured the migration of silver from AgNPs polyethylene packaging based on titanium dioxide (TiO_2) into Penaeus semisulcatus by a titration comparison within the other migrations, and found that titration had a superior sensitivity compared to the other migration methods in determining the residues of nanoparticles ($p < 0.05$) [50]. Hannon et al. determined the silver release from an experimental AgNPs spray coated on the surface of polyester and LDPE packaging material into milk [51]. The test of coating process suggested the process modification has the potential to reduce migration. Becaro et al. evaluated the genotoxic and cytotoxic effects of AgNPs (size range between 2 and 8 nm) on root meristematic cells of Allium cepa (A. cepa) [52]. The related studies often concentrate on the inhibition of growth of harmful bacteria. Interestingly, Mikiciuk et al. reported that the concentration and type of AgNPs solutions had an important effect on probiotic bacteria [53]. These bacteria were isolated from fermented milk products beneficial for the digestive system, including *Lactobacillus* acidophilus LA-5, *Bifidobacterium* animalis subsp. lactis BB-12 and *Streptococcus* thermophilus ST-Y31, which deserves great public attention.

2.2. Zinc Oxide Nanoparticles

Zinc oxide (ZnO) has attracted great interest worldwide because of its excellent properties, particularly resulting from the realization of the development of nanomaterials. Considerable studies of ZnO-NPs have been triggered on the production of nanoparticles using different synthesis methods and on their future applications, attributed to their high luminescent efficiency with a large exciton binding energy (60 meV) and a wide band gap (3.36 eV) [54]. ZnO-NPs usually act as antimicrobial and UV-protective agents used in the food packaging area. The increasing focus on ZnO-NPs drives the innovative development of synthesis methods of nanoparticles and their functions (Table 1).

Table 1. Application of Zn/ZnO nanoparticles [54].

Field of Application	Example
Biology and medicine	Bio-imaging
	Drug and gene delivery
	Antitumor and antimicrobial activity
Cosmetic industry	UV filters in sunscreens
	Mineral cosmetics
Manufacturing and materials	Antimicrobial food packaging
	Protection from exposure to UV rays
	Antimicrobial textiles
Energy and electronics	Chemical sensors based on zinc oxide
	Low cost solar cells
	Nano-generator power sensors based on ZnO nanowires

2.2.1. Synthesis Methods

Synthesis methods of zinc oxide have been developing rapidly. Because the synthesis approach determines the properties of nanomaterial, the selection of synthetic methods is a crucial step in the engineering of ZnO-NPs for a decided utilization. In recent decades, three main approaches have been used for forming ZnO-NPs: physical, chemical and biological methods. Among them, the casting method followed by solvent evaporation is the most common method used in preparation of ZnO nanocomposites with different morphologies. Rokbani et al. reported a synthesis method using a combination of ultrasound stimulations and autoclaving to prepare electrospun nanofibers of mesoporous silica doped with ZnO-NPs [55]. Jafarzadeh et al. used the solvent casting method to prepare nanocomposite films of nano-kaolin and ZnO nanorod (ZnO-nr) complex embedded into semolina film matrices [56]. Youssef et al. prepared a novel bionanocomposites packaging material using carboxymethyl cellulose (CMC), chitosan (CH) and ZnO-NPs by the casting method [18]. Salarbashi et al. developed a soluble soybean polysaccharide (SSPS) nanocomposite incorporating ZnO-NPs using a solvent-casting method [57]. Shahmohammadi and Almasi obtained bacterial cellulose-based monolayers and multilayer films with 5 wt% ZnO-NPs incorporated by using ultrasound irradiation (40 kHz) during ZnO-BC nanocomposites preparation [58]. Akbariazam et al. prepared a novel bionanocomposite of soluble soybean polysaccharide (SSPS) and nanorod-rich ZnO by the casting method [59].

2.2.2. Physical and Chemical Properties and Biological Activity

Compared with traditional antimicrobial agents, metal oxide nanoparticles show higher stability under extreme conditions with antimicrobial activity at low concentrations, and are considered to be non-toxic for humans [54]. Among these metal oxide materials, ZnO-NP is a strong antimicrobial agent [60]. ZnO-NPs exhibited diverse morphologies and showed robust inhibition against growth of broad-spectrum bacterial species. Mizielinska et al. studied the effect of UV on the mechanical properties and the antimicrobial activity against tested microorganisms of PLA/ZnO-NPs films [61]. They found that a decrease in Q-SUN irradiation to the antimicrobial activity of films with ZnO-NPs against *B. cereus*, whereas Q-UV and UV-A irradiation showed no effect on the mechanical properties of developed nanomaterial. Kotharangannagari and Krishnan studied the shape memory properties of novel biodegradable nanocomposites made of starch, polypropylene glycol (PPG), lysine and ZnO-NPs [62]. The results showed shape memory properties in the prepared nanocomposites by treating the sample at 25 °C and then at 55 °C. Furthermore, the mechanical properties showed an increase with increasing of ZnO-NPs content. Babaei-Ghazvini et al. investigated the UV-protective property of the prepared biodegradable nanocomposite films incorporated by starch, kefiran and ZnO-NPs, with a function of ZnO-NPs at different contents (1, 3, and 5 wt%) [63]. The tensile strength and Young's modulus of the specimens were measured and found that they were increased with Zn content up to 3 wt%, whereas elongation at break of the material was decreased. Besides, it is indicated that an increase of Tm following with Zn content increased thermal properties. Mizielinska et al. reported a test of change in adhesiveness of fish samples stored in fillets in active coating boxes [64]. The result showed a decrease of adhesiveness of the fish sample when stored in an active container. Besides, it was found that packaging materials containing ZnO-NPs were more active against cells of psychotropic and mesophilic bacteria than the coatings with polylysine after 144 h and 72 h of storage. Calderon et al. developed a Zn-ZnO core-shell structure and explored the oxidation capability of carbon supported Zn nanostructures used as oxygen scavenging materials activated by the relative humidity in the environment [65].

2.2.3. Applications in Food Systems

ZnO-NPs are recognized as inexpensive with potential antimicrobial properties. So the applications of ZnO-NPs packaging in food systems concentrate on its antibacterial effect, and they are used to prolong the fresh food products' shelf-life. Youssef et al. used an innovative carboxymethyl cellulose/chitosan/ZnO bionanocomposite film to enhance the shelf-life of Egyptian soft white cheese [18]. Mizielinska et al. compared the impacts of material containing polylysine or ZnO-NPs on the texture of Cod fillets, and found a lowest water loss when the sample was packed with ZnO-NPs, and an increase in the adhesiveness of the fish samples stored in boxes without active coatings, indicating that ZnO-NPs prevented the adhesiveness of food products [64]. Li et al. estimated the influences of ZnO-NPs incorporation into PLA films on the quality of fresh-cut apples [66]. It was found that the most weight loss was observed in nano-blend packaging films compared to the PLA film at the end of storage; however, packaging nanomaterial provided a better maintenance of firmness, color, sensory quality and total phenolic content. It also exhibited a strong inhibition against the growth of microorganisms. Beak et al. proposed that the synthesized olive flounder bone gelatin/ZnO-NPs film showed antimicrobial activity against *L. monocytogenes* contamination on spinach but with no effect on its quality, mainly including color and vitamin C content [67]. Suo et al. found that ZnO-NPs-coated packaging films increased the occurrence of microorganism injury, which was helpful to control pork meat in cold storage [68]. Al-Shabib et al. prepared *Nigella sativa* seed extract-zinc nanostructures (NS-ZnNPs) material and found that NS-ZnNPs showed inhibition effects on the biofilm formation of four food pathogens including *C. violaceum* 12472, *L. monocytogenes*, *E. coli*, *PAO1*, at their sub-inhibitory concentrations [69].

2.2.4. Safety Assessments

ZnO-NPs are utilized as active materials in food packaging, which might bring a potential risk for consumers contacting with this material. This nanoparticle has been demonstrated in in vivo studies that they can access organs through different pathways such as ingestion, inhalation, and parenteral routes [54]. Ansar et al. suggested that hesperidin augmented antioxidant defense with antiphlogistic reaction against neurotoxicity induced by ZnO-NPs, and the enzyme activity enhanced the antioxidant potential to reduce oxidative stress [70]. Senapati et al. evaluated the immune-toxicity of ZnO-NPs in different ages of BALB/c mice after sub-acute exposure, and found that the aged mice were more susceptible to ZnO-NPs-induced immune-toxicity [71]. Meanwhile, information on the amount of ZnO-NPs contained in food packaging and the impacts of their exposure on intestinal function are still insufficient. Moreno-Olivas et al. found that the amount of zinc present in the food was about 100 times higher than the recommended dietary allowance [72]. The effects of ZnO-NP exposure to the small intestine composed of Caco-2 and HT29-MTX cells was investigated in an in vitro model. It was found that Fe transport and glucose transport following ZnO NPs exposure were 75% decreased and 30% decreased, respectively. Also, the ZnO-NPs affected the microvilli of the intestinal cells. Zhang et al. reported the fate of the packaging material of ZnO-NPs on the coating layer incorporated into PLA-coated paper entering into paper recycling processes [73]. The results of mass balance indicated that 86–91% ZnO-NPs ended up in the material stream, mostly incorporated into the polymer coating; however, 7–16% nanoparticles completed in the desired material stream. Furthermore, the nano-coating showed positive impacts on the quality of recovered fiber. Chia and Leong made a surface modification to decrease the toxicity of ZnO-NPs by silica coating and found a significant decrease on the dissolution of ZnO-NPs [74]. They suggested that the coating offered a possible solution to enhance the biocompatibility of ZnO-NPs, which could broaden the applications such as antibacterial agent in food packaging.

2.3. Copper-Based Nanoparticles

Copper-based nanoparticles mainly include copper nanoparticles (CuNPs) and copper oxide nanoparticles (CuO-NPs). Most studies focusing on CuO NPs suggest that this material is one of the most-extensively studied metal oxide nanoparticles. The antimicrobial activity is its important feature, thus this material can be used to reduce the growth of bacteria, viruses and fungi. The nano-sized CuO-NPs were allowed to interact with the cell membrane due to their enormous surface area, and then showed an increased antimicrobial effect [26]. CuO-NPs have been applied intensively in chemical engineering and food and biomedical areas, and used as gas sensors, catalysts, water disinfectants, polymer reinforcing agents, and as a material of food packaging, semiconductors, magnetic storage media, solar cells field, emission devices and so on [75]. Consequently, antibacterial activity of CuO-NPs has been widely utilized in the fields of food packaging materials, polymer nanocomposites and water purification.

2.3.1. Synthesis Methods

CuO-NPs have potentials for forming antimicrobial nanohybrids. Almasi et al. claimed that whether the polymer substrate has already exhibited antimicrobial activity or not, the incorporated CuO-NPs could further increase the activity of the two components contained in the nanocomposite [26]. They have fabricated a novel nanocomposite incorporation into CuO-NPs, bacterial cellulose nanofibers and chitosan nanofibers by a chemical precipitation method. Gu et al. introduced a green, facile and low cost biosynthesis of monoclinic CuO-NPs based on an ultrasound method by using the extracts of *Cystoseira trinodis* as an eco-friendly material [76]. Eivazihollagh et al. reported a facile method to synthesize spherical CuNPs in situ templated by a gelled cellulose II matrix under the alkaline aqueous conditions [77]. No more than 20 min, the nanocomposite material was harvested in a one-pot reaction. Castro Mayorga et al. prepared an active biodegradable nanocomposites of poly(3-hydroxybutyrate-co-3-hydroxyvalerate) (PHBV) melt mixed with CuO-NPs in bilayer structures [78]. This bilayer-structural material was made of an active electrospun fibers mat embedded by PHBV18 (18% valerate) and CuO-NPs, and coating onto a bottom layer of concentration molded PHBV3 (3% mol valerate). Gautam and Mishra [79] synthesized Cu-NPs material compositing edible bilayer pocket prepared by heat and NaBH$_4$ treated methods to form a heat-sealable casein protein layer laminated with sodium alginate-pectin.

2.3.2. Physical and Chemical Properties and Biological Activity

The properties of the CuO-NPs depend on the synthesis method and they are very important for their applications in various areas, such as food packaging research, which rely on their biological activity. Although the specific mechanism of the antimicrobial effect of CuO nanoparticles is little known, their antimicrobial actions on bacterial cells have been proposed [75]. Beigmohammadi et al. determined the antimicrobial LDPE packaging films incorporating AgNPs, CuO-NPs and ZnO-NPs in testing of coliform amounts of ultra-filtrated cheese [80]. The results showed the number of surviving coliform bacteria declined to 4.21 log CFU/g after storing for 4 weeks at 4 ± 0.5 °C for all three treatments. Almasi et al. found that the antimicrobial activity of CuO-NPs against both Gram-negative and Gram-positive bacteria was inhibited after attachment to bacterial cellulose nanofibers; however, a synergistic action presented between chitosan nanofibers and CuO-NPs on the antimicrobial activity was reinforced [26]. Shankar et al. evaluated the water vapor permeability, barrier property, UV and thermal stability, and antimicrobial activity of the nanocomposite films [81]. The types of polymers used decided the surface morphology of films. The results showed that the addition of CuO-NPs increased the above-mentioned properties, and the films showed antimicrobial activity against *Listeria monocytogenes* and *E. coli*.

2.3.3. Applications in Food Systems

Nanomaterials with various characteristics generated from many polymers constructing copper-based nanocomposites can be used in a variety of applications. The antimicrobial activity of copper-based nanocomposites reveals applications in engineering food packaging, textile industry, medical devices and water decontamination. Some applications in food packaging over the recent three years are presented in Table 2. However, the actual applications in real food samples were little. Gautam and Mishra synthesized copper-based nanocomposite incorporating a pectin layer to enhance its antimicrobial activities [79]. It was applied in packaging coconut oil and then the oxidative stability of oil was investigated during storage. The results showed that its thermal stability was enhanced due to the nanocomposite addition, and antimicrobial activity of heat-treated nanocomposite film was increased compared to the NaBH$_4$-treated NPs film against the growth of *E. coli*. Li et al. prepared chitosan/soy protein isolates nanocomposite film reinforced by Cu nanoclusters, and this material showed the improved elongation at break and tensile strength, and higher water contact angle and degradation temperature and decreased water vapor permeation [82]. Lomate et al. developed an LDPE/Cu nanocomposite film in food packaging to extend the shelf-life of peda, which is an Indian sweet dairy product [83].

Table 2. Main applications of copper-polymer nanocomposites [84].

Polymer Matrix	Microorganism	Food Packaging Application
Cellulose	*S. cerevisiae*	Fruit juices
Hydroxypropyl methylcellulose	*S. epidermis,* *Streptococcus A.,* *E. faecalis,* *B. cereus,* *P. aeruginosa,* *Salmonella,* *Staphylococcus aureus*	Meat
Polylactic acid	*Pseudomonas* spp.	*Not mentioned*
Agar	*L. monocytogenes,* *E. coli*	*Not mentioned*
High density polyethylene	*E. coli DHSα*	*Not mentioned*

2.3.4. Safety Assessments

Although copper-based nanocomposites have been applied for diverse purposes, CuO-NPs and copper ions can be released from the packaging materials into the food systems. Little is known on the toxicity of copper-based nanocomposites and more attentions have been concentrated on CuO NPs [84].

2.4. TiO$_2$ Nanoparticles

Titanium dioxide nanoparticles (TiO$_2$-NPs), as among the most explored materials, are considered as valuable metal oxide nanomaterials with thermostability and inertia. The material also has the ability to modify the properties of biodegradable films. Besides, this material has many advantages such as being cheap, nontoxic, and photo-stable. It has been emerging as a superior photo-catalyst material for energy and environmental fields, such as air and water purification, antimicrobial, self-cleaning surfaces and water splitting [85]. TiO$_2$-NPs and their applications in the food packaging area have attracted extensive attentions attributed to their antimicrobial activity.

2.4.1. Synthesis Methods

TiO$_2$-NPs tend to aggregate, which would possibly influence the function of film properties. Changing the surface properties of TiO$_2$-NPs is an alternative method to solve this problem. In a recent study, several ionic surfactants have been used to non-covalently attach to the surface of TiO$_2$-NPs, and more innovational nanocomposites have been synthesized to enhance the TiO$_2$-NPs dispersion behavior [86]. He et al. fabricated biodegradable fish skin gelatin-TiO$_2$ nanocomposite films by a solvent evaporation method [87]. Li et al. proposed a facile and green method to synthesize super-hydrophobic paper by using a layer-by-layer deposition of TiO$_2$-NPs/sodium alginate multilayers onto a paper surface and then with colloidal carnauba wax adsorbed [88]. Lopez de Dicastillo et al. obtained TiO$_2$ nanotubes by a deposition process of atomic layer covering the electrospun polyvinyl alcohol (PVA) nanofibers at different temperatures to obtain antibacterial nanostructures with a relative high selective area [89]. Nesic et al. prepared eco-environmental pectin-TiO$_2$ nanocomposite aerogels by a sol-gel process and followed by drying under the supercritical conditions [90]. Firstly, pectin was dissolved in water and proper amount of TiO$_2$ colloid was added; then crosslinking reaction was induced in the presence of zinc ions and tert-butanol [90]. Finally, the gels were subjected to solvent exchange and supercritical CO$_2$ drying.

2.4.2. Physical and Chemical Properties and Biological Activity

Considerable studies indicated that addition of TiO$_2$-NPs had promoted the suitability of developed films applied in food packaging. The functional properties of these nanomaterials can be tailored by synthesizing composites that combine the properties of the individual component to achieve synergistic effects. Xing et al. investigated the impact of TiO$_2$-NPs on the physical and antimicrobial capabilities of polyethylene (PE)-based films. They found that the antimicrobial activity of the films was attributed to the biocidal action of TiO$_2$-NPs against bacteria [91]. Roilo et al. prepared bilayer membranes for food packaging applications by depositing TiO$_2$-NPs on PLA substrates, cellulose nanofibers and nanocomposite coatings, and found that the addition of TiO$_2$-NPs reduced the penetrant diffusivity but did not affect gas barrier performances, as well as slightly decreased the optical transparency [92]. Oleyaei et al. developed TiO$_2$-NPs (0.5, 1 and 2 wt%) incorporated into potato starch films [93]. It was found that TiO$_2$-NPs enhanced the optical transparency, and slightly increased the tensile strength and contact angle, and significantly declined the water vapor permeability properties, and decreased the elongation at break of the film. Goudarzi et al. produced the eco-friendly starch and TiO$_2$-NPs bio-composites at different TiO$_2$-NPs contents (1, 3, and 5 wt%), and investigated the mechanical, physical, water-vapor permeability (WVP), thermal properties and as well as UV transmittance of the synthesized nanomaterial [94]. They found that hydrophobicity increased, and elongation at break and tensile energy to break increased, while tensile strength, WVP and Young's modulus reduced with increasing TiO$_2$ content. Abdel Rehim et al. prepared the photo-catalytic paper sheets by adding different ratios of TiO$_2$-NPs/sodium alginate nanocomposite [95]. It was found that biopolymer of sodium alginate reduced the negative effect of the photo-catalyst on paper fibers and increased the adhesion of TiO$_2$-NPs to them.

2.4.3. Applications in Food Systems

TiO$_2$-NPs are antimicrobial agents. When they are irradiated with UV light, many reactive oxygen species are produced that have the ability to kill microorganisms. Moreover, as nano-additives, TiO$_2$ can improve the mechanical properties of polymer nanocomposites [92]. Mihaly-Cozmuta et al. synthesized three active papers based on cellulose, mainly containing TiO$_2$, Ag-TiO$_2$ and Ag-TiO$_2$-zeolite nanocomposites (P-TiO$_2$, P-Ag-TiO$_2$, P-Ag-TiO$_2$-Z), which aimed at being applied in bread packaging [96]. The efficiency in the bread storage was compared in terms of nutritional parameters (proteins, total fat and carbohydrates), acidity, and change of molds and yeasts. Li et al.

prepared a novel nano-TiO_2-LDPE (NT-LDPE) packaging. They investigated the effects of NT-LDPE material packaging on the antioxidant activity and quality of strawberries [97].

2.4.4. Safety Assessments

Nanoparticles exhibited an increased surface-to-mass ratio enhancing the reactivity. Moreover, nanoparticles displayed an increased tendency to penetrate the cell membranes and consequently having the potential to transfer through the biological barriers. So far, the health effects of TiO_2-NPs have been explored basically on their uptake by inhalation. It was concluded by the International Agency for Research on Cancer (IARC) that decided based epidemiological studies to assess whether TiO_2 dust causing human cancers was inadequate. Evidence for carcinogenicity in experimental animals was sufficient, which was conducted on account of the induction of respiratory tract tumors in rats after prolonged inhalation. Accordingly, TiO_2-NPs is classified as a Group 2B carcinogen by the IARC [98]. Based on the extensive food-related uses, an increasing attention has been drawn to the risk assessment of TiO_2-NPs applied in food packaging. Ozgur et al. evaluated the effect of different amounts of TiO_2-NPs (0.01, 0.1, 0.5, 1, 10 and 50 mg/L) in vitro at 4 °C for 3 h on sperm cell kinematics with the velocities of Rainbow trout [99]. Additionally, oxidative stress markers (superoxide dismutase (SOD) and total glutathione (TGSH)) of the sperm cells were tested after their exposure to TiO_2-NPs. The results revealed that a statistical significance ($p < 0.05$) presented in the velocities of sperm cells. When concentration of TiO_2-NPs reached at 10 mg/L, an increased activity of TGSH and SOD ($p < 0.05$) levels were found. Salarbashi et al. prepared biodegradable SSPS nanocomposites consisting of varying ratios of SSPS and TiO_2-NPs, and found that TiO_2-NPs existed in plasma membranes of epithelial cell lines after a 10-day exposure to a number of free nanomaterials [100]. However, anti-cancerous and pro-cancerous activities were not determined because this nanomaterial denoted their neutrality in regards to cancer inhibition or promotion in gastrointestinal tracts. Jo et al. evaluated the interactions between TiO_2-NPs and biomolecules including albumin and glucose [101]. They investigated that those biomolecules altered the physical and chemical properties as well as the consequence regarding TiO_2-NPs under physiological conditions. It was found that oral absorption of food grade TiO_2-NPs was slightly higher compared to general grade TiO_2-NPs; however, these nanoparticles were excreted through the feces. Besides, the biokinetics of food grade TiO_2-NPs were extremely relied on their interaction with biomolecules.

2.5. Other Metal Oxide and Nonmetal Oxide Nanomaterials

In addition to the metal oxides mentioned above, several other metal oxide and nonmetal oxide nanomaterials also showed an increased potential using as packaging materials, such as MgO-NPs, Fe_3O_4-NPs and iron-based nanoparticles, as well as SiO_2-NPs [102–105].

MgO naturally exists as a renewable, colorless, crystalline mineral and is economically produced on a large scale. The use of MgO has been recognized as generally safe even in the food applications by the U.S. Food and Drug Administration (FDA). Swaroop and Shukla produced films by incorporating MgO-NPs in PLA polymer through a solvent casting method, and found 2 wt% amount of MgO-NPs in PLA films exhibited the most observed improvement in the oxygen barrier and tensile strength properties, as well as a superior antibacterial efficacy; whereas, nearly a 25% negative effect was found on water vapor barrier properties [102].

Ren et al. synthesized inorganic materials of magnetic ferroferric oxide nano-particles in-situ coating on graphene oxide nanosheets (Fe_3O_4@GO) as fillers and then were used to fabricate a PVA nanocomposite film [106]. This material showed a superior barrier capability regarded as a better choice compared to the traditional aluminum films. Shariatinia and Fazli prepared a thickness of 0.13–0.2 mm nanocomposite film made of starch, chitosan, cyclophosphamide, glycerin and Fe_3O_4-NPs [103]. Khalaj et al. prepared nanocomposites of the nano-clay containing iron nanoparticles (Fe-NPs)-polypropylene (PP) by a melt interaction. They investigated the morphological, mechanical, gas barrier and thermal properties [107].

The activities of SiO_2-NPs are related to their average particle size, biocompatibility, high surface area, stability, low toxicity, bad thermal conductivity and supreme insulation [108]. Mallakpour and Nazari developed a facile and fast method to synthesize polymer-based nanocomposite films of PVA and SiO_2-NPs coating on bovine serum albumin (PVA/SiO_2@BSA) using a casting method assisted by sonication [108]. Guo et al. investigated the impacts of realistic doses in physiological terms of SiO_2-NP on gastrointestinal function and health, based on an in vitro model composed of HT29-MTX and Caco-2 co-cultures representing goblet and absorptive cells, respectively [109]. The results showed that the exposure of SiO_2-NPs was harmful to the brush border membrane and that exposure to the physiologically relevant doses of well-characterized SiO_2-NP for acute (4 hours) and chronic (5 days) time periods eventually led to adverse effects in cells.

2.6. Nano-Clay and Silicate Nanoparticles

At the present time, nano-clay has approximately 70% market value all over the world, meaning that it is the most commercially applied nanomaterial. There are various kinds of clay minerals according to their structures and chemistries as well as sources. Based on the layered structures, these materials are categorized into four major classes, i.e., chlorite, montmorillonite (MMT)/smectite, illite and kaolinite [110]. MMT is recognized as the most commonly used in the preparation of nanocomposites among these clays. The widest acceptability in layered clay is obvious since it has high surface reactivity and surface area [27]. As a result, many studies demonstrated that natural biopolymer-layered silicate nanocomposites significantly improved properties in packaging. In spite of this, there were fewer studies of orgnoclays as nanomaterials in food packaging compared to other nano-encapsulation systems [111].

2.6.1. Synthesis Methods

Halloysite nanotubes (HNTs), as a kind of natural nanomaterials belonging to kaolinite, have a hollow tubular-like structure within the inner and outer diameters of 15 nm and 50 nm, respectively. Due to this tubular shape, HNTs have a capability of being loaded by various materials, which have been developed as functional nanocapsules [112]. Biddeci et al. prepared a functional biopolymer film by filling a pectin matrix modified with HNTs containing peppermint oil, where HNT surfaces were functionalized with cucurbit uril molecules with the aim to enhance the affinity of the nanofiller towards peppermint oil [16]. Pereira et al. prepared lycopene and MMT-NPs in whey protein concentrate films using the casting/evaporation method [113]. Recently, several agricultural processing wastes have been used to synthesize the nanocomposites as raw materials. Orsuwan and Sothornvit developed a biopolymer film incorporated with banana starch nanoparticles (BSNs) and MMT-NPs, where the BSNs was fabricated using miniemulsion cross-linking to make an enhanced agent [114]. Oliveira et al. used the pectin extracted from pomegranate peels to prepare films with the some amounts of MMT-NPs as reinforcement nanomaterial [115]. Zahedi et al. investigated a novel casting method to fabricate a carboxymethyl cellulose (CMC)-based nanocomposite films containing MMT (5 wt%) and ZnO-NPs (1, 2, 3 and 4 wt%) and found addition of ZnO-NPs enhanced the UV-light blocking (from 60% to 99%) of single-layer nano-clay [116].

2.6.2. Physical and Chemical Properties and Biological Activity

Nano-clays, especially MMT, act as crucial fillers in the biodegradation of nanomaterials when incorporated with a polymer, because they are toxin-free, environmentally friendly and safe to be used in food packaging. Besides, their activities to reduce permeability of gases and improve mechanical properties have been confirmed for polymer nanomaterials. Kim et al. investigated a potential application of multilayer packaging films for packing food containing waterborne content, which were prepared by dry laminating commercially available PVA/vermiculite nanocomposites [117]. They found a reversible regression of the barrier properties of oxygen presented in the prepared films. Pereira et al. characterized the structural and mechanical properties of lycopene/MMT/whey protein concentrate films and found that MMT at the amount of 20 g/kg in the polymeric matrix increased both thermal and mechanical properties [113]. Besides the red coloring ability, lycopene showed no effects on detectable interference in the physical or structural properties. Beigzadeh Ghelejlu et al. prepared nano-clay nanocomposite/chitosan active films incorporated with three levels of Silybum marianum L. extract (SME) (0.5% *v/v*, 1% *v/v* and 1.5% *v/v*) and MMT (1, 3 and 5 wt% of chitosan) [19]. The results indicated that the addition of SME and MMT improved the antioxidant properties of the films, but decreased the solubility and WVP and influenced the optical and mechanical properties of films. Notably, plant essential oils have been encapsulated into nano-clay or MMT-NPs to improve the antioxidant and antimicrobial activities of composite materials applied in packaging system, including peppermint, thyme and cinnamon [16,17,118]. Khalaj et al. found that in the prepared nanocomposite of Fe/MMT/PP-NPs, the intercalation and exfoliation of the clay were affected reversely after the addition of PP-NPs to some extent [107]. Furthermore, certain homogeneity of uniform distribution of MMT and PP-NPs was observed through TEM and SEM. The melting temperatures increased with clay concentration; however, crystallization temperature and crystallinity decreased with the clay concentration with NPs compensating the effect of clay. Nano-clays also have prospects in active and intelligent food nano-packaging. Gutiérrez et al. developed a nano-clay of MMT containing blueberry extract [111]. They revealed that according to a shift between flavylium and quinoidal forms of anthocyanins in blueberries, the color was changed following the pH of the system. Thus, addition of blueberry extract could modify the structure of MMT to form novel nano-clays with more active properties.

2.6.3. Applications in Food Systems

Generally, clays are low-cost, naturally occurring and eco-friendly agents and used in various applications. Clay minerals are used in the fields of agriculture, geology, engineering, construction and process industries [27]. Peter et al. investigated the chemical and microbiological characteristics of white bread during the storage in paper packaging modified with Ag/TiO$_2$-SiO$_2$-NPs [119]. The results showed good water retention and prolonged shelf-life of bread for 2 days compared to the unmodified packaging. Nalcabasmaz et al. developed nanocomposite materials containing 1% nano-clay and 5% poly-beta-pinene (P beta P) [120]. They tested the material for packaging sliced salami. The packaged food sample used nanofilms and multi-layered film under different conditions of vacuum, modified atmosphere packaging of 50% CO$_2$ and 50% N$_2$ and air, and both stored at 4 °C for 90 days. It was found that the moisture content and hardness showed no significant changes during storage. The sliced salami stored under vacuum and high CO$_2$ using the multilayer material displayed the longest storage time of 75 days. Kim et al. developed insect-proof HNTs material, which were applied to a LDPE-based film to control *Plodia interpunctella* (Indian mealmoth) from infesting the food [121]. Peighambardoust et al. prepared LDPE-based Films incorporating with organic clay nanoparticle including cloisite 30B, cloisite 20A and cloisite 15A for packaging to decrease the growth of coliform bacteria in ultra-filtrated cheese [122]. The developed films exhibited a decrease of coliform load to 2.05 log CFU/g at the optimum condition, which was corresponding to Japanese industrial standard (JIS Z 2801:2000). Echeverria et al. evaluated the future application of active nanocomposite films based on soy protein isolate-MMT loaded with clove essential oil to preserve the muscle fillets of Bluefin tuna stored in refrigerator [123]. They further analyzed the possibility of clay in packaging diffusing to the food system. Clay inclined to release the clove oil by extending its antimicrobial activity (especially against *Pseudomonas* spp.) and enhancing antioxidant activity. There were no observed metals (Si and Al) of clay diffused to the muscle of fish. Guimaraes et al. evaluated fresh-cut carrots (FCC) coated by MMT-NPs subjected to packaging of passive modified atmosphere [124]. The use of starch nanoparticles incorporated into coating film together with a modified atmosphere led to the enhanced total antioxidant activity, volatiles, and organic acids maintaining of FCC. Junqueira-Goncalves et al. evaluated the effect of addition of MMT-NPs to a lacto-biopolymer coating [125]. They found that the material could improve its water vapor barrier and reduce weight loss, as well as oxygen uptake and the release of carbon dioxide, and improve fruit firmness and reduce mold and yeast load, at last prolong the shelf-life of coated strawberries.

2.6.4. Safety Assessments

With nano-clay or MMT-NPs materials attracting more and more attentions worldwide, analyses of risks of these nanomaterials to the lung health of exposed workers have been emerging. Besides, present studies aiming to demonstrate the toxicological actions of nano-clay showed that the structure had resulted in the promotion of cellular uptake and interactions [126]. Han et al. presented a study on the degradation and release of nano-clay-loaded LDPE composite for food packaging [127]. It was found that the toxicity of released nano-clay particles from nano-clay particle-embedded LDPE composites to A594 adenocarcinomic human alveolar basal epithelial cells was degraded. Wagner et al. investigated the potential of Cloisite 30B and Cloisite Na$^+$ and their thermally degraded byproducts and then induced toxicity in the model of lung epithelial cells of human [126]. Analysis of byproduct physical and chemical properties suggested changes happened in structures and functions. Echegoyen et al. investigated the migration of nano-clay from food packaging materials to food samples [128]. The results showed that Al-NPs of different sizes and morphologies could migrate into food stimulants with different food stimulants (acetic acid 3% and ethanol 10%), temperatures and times (70 °C for 2 h and 40 °C for 10 days) from two commercialized LDPE-based nanocomposite bags analyzed by ICP-MS.

3. Organic Biopolymer-Based Nanomaterials Applied in Food Packaging

The concept of a bio-based economy is gradually receiving attentions from scientific, societal, and economic aspects, and there is a great deal of driving force to develop strategies for this purpose [129]. The inspiration of producing biopolymer-based materials is to utilize renewable organic sources, including polysaccharides and proteins, aiming at replacing non-renewable fossil sources. There are various organic nanomaterials applied in food packaging, mainly divided into three categories: polymer-based plastics, polysaccharide-based and protein-based nanomaterials, which provide biopolymer matrix for nanocomposite materials.

3.1. Polymer-Based Nanomaterials

Traditionally, most plastic packaging materials are made from petroleum-based polymers, mainly containing the commodity polystyrene (PS) and polyethylene (PE), which are recognized as not being environmentally friendly. With the development of new polymer materials, the biodegradable polymer-based plastics can provide a viable alternative, such as PVA [130–134], polylactic acid (PLA) [135–139], poly(3-hydroxybutyrate) (PHB) [140,141] and poly(3-hydroxybutyrate-co-3-hydroxyvalerate) (PHBV) [78,142] and their biopolymer blends.

3.1.1. PVA

PVA, basically made from polyvinyl acetate through hydrolysis, is easily degraded by biological organisms in water. It has been extensively incorporated into other polymer-based compounds to increase the mechanical properties attributed to its hydrophilic properties and compatible structure, including mechanical performance, solvent resistance, biocompatibility and high hydrophilicity [143]. Yang et al. prepared chitosan/PVA hydrogels containing lignin nanoparticles (LNPs) (1 wt% and 3 wt%) by a freezing–thaw procedure [130]. The study of mechanical, microstructural and thermal characterizations of the nanomaterial showed that the optimal amount of LNPs was at 1 wt%, whereas the agglomerates at higher LNP content were formed and affected the properties [144]. Sarwar et al. investigated the impact of Ag-NPs embedded into nanocellulose on the mechanical, physical and thermal properties of PVA-based nanocomposite films [131]. They found that these films had a superior antimicrobial activity against *E. coli* (DH5-alpha) and *S. aureus* (MRSA). Furthermore, the films showed no cytotoxicity effect on HepG2 and the cell viability was above 90%.

3.1.2. PLA

PLA has drawn more attention resulting from the good biodegradability and being a candidate of substitution for traditional polymers. PLA is mainly produced by condensation polymerization from lactic acid, derived from fermentation of corn, sugars, tapioca or sugarcane. Among the various biopolymers investigated, PLA exhibits key properties, including biodegradability, renewability and superior mechanical properties, crystallinity and process ability [102,145]. Aframehr et al. investigated the impact of calcium carbonate ($CaCO_3$) nanoparticles on the biodegradability and barrier properties of PLA [137]. The results showed that the barrier properties were increased by loading $CaCO_3$-NPs increasing to 5 wt%. It was also found that the gas permeability of CO_2, O_2 and N_2 were enhanced by increasing temperature but decreased by increasing feeding pressure. Vasile et al. prepared the Cu-doped ZnO powder embedded into PLA samples functionalized with Ag-NPs composites by a melt blending process [136]. The results showed an increase of the crystalline degree of PLA when the content of nanoparticle was increased from 0.5 wt% to 1.5 wt%.

3.1.3. PHBV

Polyhydroxyalkanoates (PHAs) have been gradually paid attention recently as biodegradable and biocompatible thermoplastics in packaging applications. The most extensively studied polymer from the PHAs is the poly(3-hydroxybutyrate), PHB, which is partially crystalline with a high rigidity and melting temperature. To decrease the crystallinity, the copolymer obtained with the insertion of 3-hydroxyvalerate (HV) units, named as PHBV, is usually employed with improved handling properties of PHB films [78]. Zembouai et al. prepared blends of PLA and PHBV at different PLA/PHBV weight ratios (0/100, 25/75, 50/50, 75/25, 100/0) through a melt compounding process [146]. The formed blends were investigated on the mutual contributions to flammability resistance, thermal stability, rheological behavior and mechanical properties. The results revealed that increasing PLA content in PLA/PHBV blends led to improved properties, such as flammability resistance and thermal stability. Shakil et al. developed the sepiolite/PHBV nanocomposite films by using the APTES grafted sepiolite through the solution-casting method [142]. The results provided evidence that the application of biodegradable nanocomposite films would lead to a more efficient water barrier and thermal properties.

3.2. Polysaccharide-Based Nanomaterials

3.2.1. Starch-Based Nanomaterials

Aqlil et al. investigated a graphene oxide (GO)-filled starch/lignin polymer bionanocomposite [147]. They found that the amount of GO had a strong influence on the mechanical properties and could reduce water vapor permeability and moisture uptake of the prepared film. Shahbazi et al. developed starch film incorporated with multi-walled carbon nanotubes with or without hydroxylation and found that the hydrophobic character of the film was greatly improved with incorporation of a nanotube [148]. Oleyaei et al. estimated the thermal, mechanical and barrier properties of TiO_2 and montmorillonite on potato starch nanocomposite films [93]. The results showed elongation at break, tensile strength, melting point and glass transition temperature of the films were improved followed the addition of MMT and TiO_2. The visible, UVA, UVB and UVC lights transmittance and water vapor permeability decreased with the increasing amounts of TiO_2 and MMT.

3.2.2. Cellulose-Based Nanomaterials

Shankar and Rhim prepared nanocellulose material and tested the effects on the properties of agar-based composite films [149]. The crystallinity index of nanocellulose (NC, 0.71) was decreased compared to the micro-crystalline cellulose (MCC, 0.81). The results demonstrated that NC could be used as an enhanced agent for the preparation of biodegradable composites films. Pal et al. synthesized reduced graphene oxide and cellulose nanocrystal incorporated into PLA matrix through a modified Hummer's method and an acid hydrolysis [150]. They found that the mechanical property of scaffold was significantly improved. Both tensile strength (23% increase) and elongation at break were increased, which indicated the nanocomposite was ductile compared to unmodified PLA. The distinct anti-bacterial efficacy was observed to inhibit against both Gram-negative *E. coli* and Gram-positive *S. aureus* bacterial strains. Liu et al. prepared starch-based nanocomposite films improved by cellulose nanocrystals to control d-limonene permeability [151]. They found that cellulose nanocrystals amount and aspect ratio were independently controlling d-limonene permeability through film-structure regulation. Lavoine et al. simulated release and diffusion of active substances made of cellulose nanofiber coating to food packaging material through calculating in a mathematical model derived from Fickian diffusion. They found the model was validated for caffeine only [152].

3.2.3. Chitosan-Based Nanomaterials

Postnova et al. studied approaches in which monolithic hydrogels were prepared through mineralization of polysaccharide by a method of green sol-gel chemistry, compared with a method through the formation of polyelectrolyte complex [153]. It was found that both approaches were available for the preparation of films with nanoparticles and chitosan bionanocomposites. Liang et al. prepared edible chitosan films incorporated with epigallocatechin gallate nanocapsules and characterized their antioxidant properties [154]. It was found that the addition of nanocapsules to chitosan hydrochloride films improved their tensile strength, whereas the percent of elongation at break and lightness was significantly decreased. Buslovich et al. developed in situ chitosan and vanillin incorporated on packaging films, containing an aqueous/ethanol solution onto a PE surface by an ultrasonic method [155]. The results showed that increased contact surface strongly inhibited the fruit microbial spoilage.

3.3. Protein-Based Nanomaterials

3.3.1. Zein-Based Nanomaterials

Zein, a group of prolamins from corn, is a Generally Recognized as Safe (GRAS) food-grade ingredient. With the hydrophobicity of three quarters of the amino acid residues in zein, zein-based nanomaterials have low WVP compared to many other bio-based films. Moreover, zein-based nanomaterials embedded with inorganic AgNC may have advantages such as low toxicity [30]. Aytac et al. synthesized thymol (THY)/gamma-Cyclodextrin(gamma-CD) inclusion complex (IC) encapsulated electro-spun zein nano-fibrous webs (zein-THY/gamma-CD-IC-NF) as a food packaging nanomaterial and found that zein-THY/gamma-CD-IC-NF (2:1) significantly inhibited the growth of bacteria in meat samples [156]. Rouf et al. prepared nanocomposite with the addition of silicate NPs (Laponite) to zein films casting from 70% ethanol solutions [157]. The changes in the surface energy of the films were evaluated using contact angle measurements and showed an increase in surface hydrophobicity. The Young's modulus and tensile strength were increased with increasing nanoparticle concentration. The glass transition temperature was increased and WVP was decreased with only a small amount of Laponite. Oymaci and Altinkaya prepared whey protein isolate (WPI)-based films embedded into zein nanoparticles (ZNPs) coated with sodium caseinate by a casting method [158]. They found that the addition of zein NPs dramatically improved the mechanical and water vapor barrier properties of the WPI with no effect on the elongation of the films. It was also found that both the fractional free volume and hydrophilicity of the WPI films decreased. Gilbert et al. prepared a biopolymer-based composite film of hydroxypropyl methylcellulose and ZNPs [159]. The results exhibited an increase in tensile strength, a decrease in elongation, and an initial increase followed by gradual decrease in Young's moduli with increasing ZNPs.

3.3.2. Whey Protein Isolate-Based Nanomaterials

Qazanfarzadeh and Kadivar prepared WPI-based composite films with different proportions of oat husk nanocellulose (ONC) obtained from acid sulfuric hydrolysis by a solution casting method [160]. They found that the crystallinity increased after acid hydrolysis. The films prepared with 5 wt% ONC showed the highest tensile strength, Young's modulus, solubility and the lowest elongation at break and moisture content. However, WVP and film transparency were decreased with the addition of ONC. Hassannia-Kolaee et al. prepared whey protein isolate/pullulan (WPI/PUL) films having different contents of nano-SiO_2 (NS) using a casting method [161]. The results revealed tensile strength of nanocomposite films was enhanced but elongation at break was declined after increasing NS content. Moisture content, water absorption and solubility in water were improved followed as increasing content of NS and the water resistance and barrier properties of the films were also improved. Water vapor permeability of films was decreased with the increasing NS content. Zhang et al. developed a chitosan/WPI film incorporated with TiO_2-NPs and found that the nanoparticles improved the compatibility of WPI and chitosan [86]. Nanoparticle incorporation increased the whiteness of chitosan/WPI film, but decreased the transparency. The elongation at break and tensile strength of nanocomposite film were increased by 12.01% and 11.51%, respectively, whereas WVP was decreased by 7.60%.

4. Mechanistic Studies of Nanomaterials in Food Packaging

Preventing microbial growth in foods is known as a critical function of packaging to meet the challenge of preserving the quality of food products. Accordingly, antimicrobial materials in food packaging are emerging as a promising technology to fulfill the demands. With the applications of antimicrobial agents in food packaging materials, the growth of bacteria is inhibited and thus the shelf-life of food products is prolonged considerably. Antimicrobial materials are grouped into two classes: organic and inorganic materials. Chitosan-based nanoparticles and chitin-based nanoparticles are typical examples of organic materials, which have lately been widely studied. In respect to the organic antimicrobial materials, some noble metals such as Ag-NPs, Cu-NPs and Au-NPs, as well as the oxidized nanomaterials including ZnO, TiO_2 and MgO have attracted much interest because of their resistance to the rough processing conditions and enhancement of strong biocidal impacts against foodborne pathogens [25].

An illustration of reported biocidal mechanisms induced by these nanoparticles is shown in Figure 1 [25]. There are several hypotheses with respect to antimicrobial mechanistic actions of nanomaterials. There was a general consensus that nanomaterials are proved to be an ideal alternative to traditional plastics and they have also served as a potential packaging material to prolong the shelf-life of food products. Because the large surface-to-volume ratio provides more direct interaction to bacterial surfaces, these nanomaterials showed excellent antibacterial properties. Particularly, cationic nanoparticles were firmly attached to the membrane of bacteria with negatively charged outer layers by electrostatic interactions. Disruption of the cell integrity resulted in the leakage of cell contents. Nanoparticles had intrinsic antibacterial activities to refuse the microbes by mimicking natural course of killing by phagocytic cells, i.e., by producing large quantity of reactive nitrogen species (RNS) and reactive oxygen species (ROS). Besides, nanomaterials could also prevent or overcome biofilm formation. Nanoparticles especially metallic nanoparticles exerted toxic effects by enhancing the natural immunity or mimicking natural immune responses by generating a large quantity of RNS or ROS. Others possibly exerted direct killing effects maybe by directly targeting cellular proteins, DNA or lipids [162].

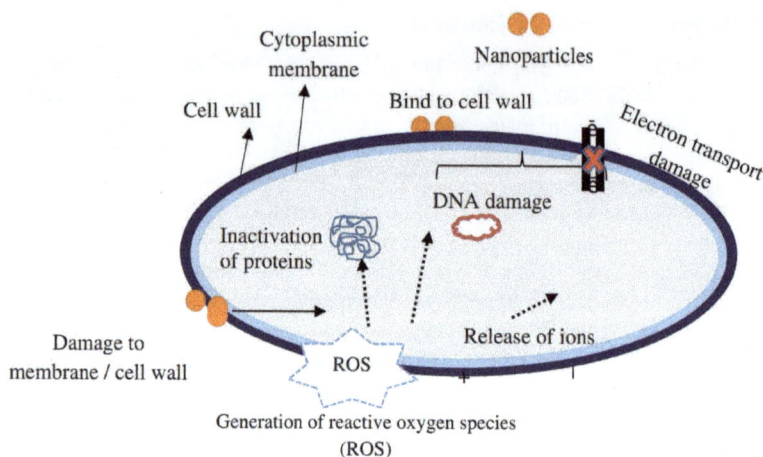

Figure 1. Schematic representation of antimicrobial mechanisms of inorganic nanoparticles. Reproduced with permission from [25]. Copyright Taylor & Francis Online, 2018.

Fardioui et al. reviewed the antimicrobial mechanisms of ZnO-NPs [54]. They found that the explicit mechanisms were still under debate, but several models were suggested as follows: (1) electrostatic interactions between cell walls and ZnO-NPs to destroy bacterial cell integrity; (2) liberation of antimicrobial Zn^{2+} ions regarding accumulation of ZnO-NPs into bacteria cells; (3) ROS formation. Shao et al. tested the electronegativity on *S. aureus* and *E. coli* surfaces after AgNPs treatment, and found a possible change in bacterial surface properties [35]. Furthermore, they observed that cell surfaces were strongly distorted after AgNPs treatment. Meanwhile, some nanoparticles were distributed in the bacterial cell surface, which indicated their direct interaction with bacteria and generation of electronic effects and enhancement of reactivity of AgNPs. El Zowalaty et al. suggested that the chitosan reacted with proteins on microbial surfaces and caused the leakage of intracellular contents [163]. It also chelated trace metal ions and disrupted the electron transport chain. Furthermore, it interfered with the formation of mRNA and proteins once inside the cell [162].

5. Certain Aspects of Concerns

In the past decades, nanomaterials in food packaging applications have been developed to enhance the barrier and mechanical properties of traditional and bio-based packaging materials, and/or to provide novel active and smart functionalities. Active and smart packaging materials deliberately incorporate active or smart components, which are intended to release or absorb substances into, onto, or from the packaged food or the surrounding environment, or to provide the intended information of their use conditions [164]. Developments of nanotechnologies are going to increasingly find utilizations in the food packaging area. However, there are gaps in our knowledge on them that put up questions to the scientific community, especially related to toxicity and ecotoxicity. Theoretically, nanoparticles have the potential to migrate to the packed food, but migration assays and risk assessment are still not conclusive [165]. Migrations into food could be considered as the process of mass transfer, in which the low-molecular mass constituents initially existed in the packaging and then released to the packed matrix. Therefore, it was considered as a diffusion process which could be described by Fick's second law. Thus, one of the most important steps during the development of novel food packaging materials is the research of the migration to investigate the probabilities of any undesirable or harmful ingredients migrating to the food products in overall and specific terms [165].

Jokar et al. summarized six main questions during the process of migration of NPs from polymer-based food contact materials to consumers in Figure 2 [166].

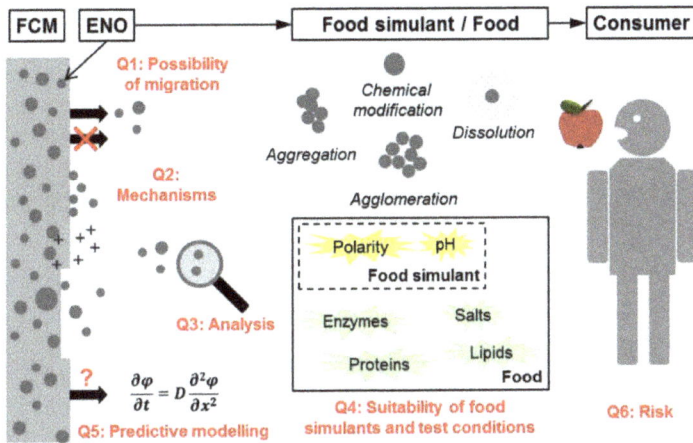

Figure 2. Overview of the six open questions about the migration of NPs from polymer-based food-contact materials identified in Jokar et al.'s review (FCM = food-contact materials, ENO = engineered nano-objects) [166].

Briefly, through the investigation, Jokar et al. found that many experimental studies had not given a conclusive answer on the possibility of migration of NPs from food packaging materials to the food products [166]. They assumed this could be partially attributed to the lack of suitable analytical methods for the detection of low quantities and small sizes of NPs. They strongly suggested that studies which concluded that no migration occurred add information about the detection limit of the measurements, including both particle mass or number concentration and particle size. Analytical techniques such as single-particle inductively coupled plasma mass spectrometry (SP-ICP-MS) have been gradually playing a role in characterizing and quantifying NPs in the food simulants extracts [167,168]. Besides, it was difficult to conclude the migrations of NPs through predictive models only considering migrations based on diffusion. Generally, three sub-processes in NP migration could be distinguished: (1) diffusion of the molecule into the polymer to food products because of a concentration gradient; (2) desorption of the molecule from the polymer and subsequent adsorption by the food at the food-packaging interface; (3) diffusion of the molecule in the food due to a concentration gradient [166]. The food stimulants and test conditions would affect the migration and ingestion actions by electrostatic interactions and chemical or mechanical decomposition and organic additives and treatment of food samples [49,169,170]. However, there was a clear lack of data on potential release mechanisms of identified NPs. The question of risk for the consumer associated with migrating NPs from food packaging probably was more complicated than other questions, because there were many influences of physio-chemical characteristics of NPs on gastrointestinal absorption, such as composition, morphology, charge, surface properties and aggregation state and food components [169]. These clearly require more exploration in the future.

In addition to the technical aspects, no regulations on nanotechnology applications have been yet established in a global context mainly because of lack of sufficient and reliable fundamental researches in regard to the safety assessment and migration characters of nanomaterials from packaging to the food system. The FDA considered that evaluations of safety, effectiveness, public health impact, or regulatory status of nanotechnology products should consider any unique properties and behaviors imparted by the application of nanotechnology [171]. The European Commission already edited statutory contents in this direction with technical guidance mentioning nanomaterials, and also recognized active and smart materials and papers to state that new technologies, which are based on engineered materials in nanoparticle sizes that exhibited physical and chemical properties are significantly different from those at a much larger scale. A risk assessment on a case-by-case basis until more is known about the novel technology is needed [172]. On this basis, and taking into account the lack of knowledge about their potential toxicity (oral exposure to nanomaterials had received less attention than the dermal or inhalation pathways [101]), the concept of functional barriers used to prevent migration of contaminants, which were not evaluated by health authorities, could not be applied in the specific case of biopolymer nanocomposites packaging. These statements notably differentiated nano-structure substances from non-nano-structure substances which were authorized for use as a functional barrier, providing that they fulfilled certain standards and the migrations remained below a given limit of detection.

6. Conclusions

The innovation of nanomaterials in the food packaging science has brought many changes in food preservation, storage, distribution and consumption. Thanks to preventing microbial growth in foods by antimicrobial activity of nanomaterials, these changes have extended the shelf-life of foods to certain degrees with better management of spoilage in food products. Furthermore, the nanotechnology provides numerous choices for cost-effective, eco-friendly, degradable and renewable packaging materials, which have been gaining more attention and acceptance to solve the ecological environment pollutions and food shortage crises by ensuring food reaches the masses. It is pertinent to note that there are some fundamental studies on toxicity and ecotoxicity, migration assays and risk assessment of nanocomposite materials still needed. In this way, it could allow nanomaterials to work better in the food packaging field.

Author Contributions: Writing: Original Draft Preparation, Y.H.; Writing: Review and Editing, Q.W.; Literature and data investigations and English corrections, L.M. and X.C.

Acknowledgments: This research was partly supported by Supported by National Nature Science Foundation of China (NSFC) (31801647), Sichuan Science and Technology Program (2018JY0194, 2018SZ0340), Key scientific research fund of Xihua University (Z1620514).

Conflicts of Interest: The authors declare no conflict of interest.

References

1. Wyser, Y.; Adams, M.; Avella, M.; Carlander, D.; Garcia, L.; Pieper, G.; Rennen, M.; Schuermans, J.; Weiss, J. Outlook and challenges of nanotechnologies for food packaging. *Packag. Technol. Sci.* **2016**, *29*, 615–648. [CrossRef]
2. Sharma, C.; Dhiman, R.; Rokana, N.; Panwar, H. Nanotechnology: An untapped resource for food packaging. *Front. Microbiol.* **2017**, *8*, 1735. [CrossRef] [PubMed]
3. Valiev, R. Materials science—Nanomaterial advantage. *Nature* **2002**, *419*, 887–889. [CrossRef] [PubMed]
4. Ariyarathna, I.R.; Rajakaruna, R.M.P.I.; Karunaratne, D.N. The rise of inorganic nanomaterial implementation in food applications. *Food Control* **2017**, *77*, 251–259. [CrossRef]
5. Pathakoti, K.; Manubolu, M.; Hwang, H.M. Nanostructures: Current uses and future applications in food science. *J. Food Drug Anal.* **2017**, *25*, 245–253. [CrossRef] [PubMed]
6. Hobson, D.W.; Roberts, S.M.; Shvedova, A.A.; Warheit, D.B.; Hinkley, G.K.; Guy, R.C. Applied nanotoxicology. *Int. J. Toxicol.* **2016**, *35*, 5–16. [CrossRef] [PubMed]
7. Siddiqi, K.S.; Husen, A.; Rao, R.A.K. A review on biosynthesis of silver nanoparticles and their biocidal properties. *J. Nanobiotechnol.* **2018**, *16*, 14. [CrossRef] [PubMed]
8. Bouwmeester, H.; van der Zande, M.; Jepson, M.A. Effects of food-borne nanomaterials on gastrointestinal tissues and microbiota. *WIREs Nanomed. Nanobiotechnol.* **2018**, *10*, e1481. [CrossRef] [PubMed]
9. Singh, T.; Shukla, S.; Kumar, P.; Wahla, V.; Bajpai, V.K. Application of nanotechnology in food science: Perception and overview. *Front. Microbiol.* **2017**, *8*, 1501. [CrossRef] [PubMed]
10. Noruzi, M. Electrospun nanofibres in agriculture and the food industry: A review. *J. Sci. Food Agric.* **2016**, *96*, 4663–4678. [CrossRef] [PubMed]
11. Han, J.W.; Ruiz-Garcia, L.; Qian, J.P.; Yang, X.T. Food packaging: A comprehensive review and future trends. *Compr. Rev. Food Sci. Food Saf.* **2018**, *17*, 860–877. [CrossRef]

12. Youssef, A.M.; El-Sayed, S.M. Bionanocomposites materials for food packaging applications: Concepts and future outlook. *Carbohydr. Polym.* **2018**, *193*, 19–27. [CrossRef] [PubMed]
13. Kuswandi, B. Environmental friendly food nano-packaging. *Environ. Chem. Lett.* **2017**, *15*, 205–221. [CrossRef]
14. Liu, Y.; Wang, S.; Zhang, R.; Lan, W.; Qin, W. Development of poly(lactic acid)/chitosan fibers loaded with essential oil for antimicrobial applications. *Nanomaterials* **2017**, *7*, 194. [CrossRef] [PubMed]
15. Yu, H.Y.; Yang, X.Y.; Lu, F.F.; Chen, G.Y.; Yao, J.M. Fabrication of multifunctional cellulose nanocrystals/poly(lactic acid) nanocomposites with silver nanoparticles by spraying method. *Carbohydr. Polym.* **2016**, *140*, 209–219. [CrossRef] [PubMed]
16. Biddeci, G.; Cavallaro, G.; Di Blasi, F.; Lazzara, G.; Massaro, M.; Milioto, S.; Parisi, F.; Riela, S.; Spinelli, G. Halloysite nanotubes loaded with peppermint essential oil as filler for functional biopolymer film. *Carbohydr. Polym.* **2016**, *152*, 548–557. [CrossRef] [PubMed]
17. Iamareerat, B.; Singh, M.; Sadiq, M.B.; Anal, A.K. Reinforced cassava starch based edible film incorporated with essential oil and sodium bentonite nanoclay as food packaging material. *J. Food Sci. Technol.* **2018**, *55*, 1953–1959. [CrossRef] [PubMed]
18. Youssef, A.M.; El-Sayed, S.M.; El-Sayed, H.S.; Salama, H.H.; Dufresne, A. Enhancement of egyptian soft white cheese shelf life using a novel chitosan/carboxymethyl cellulose/zinc oxide bionanocomposite film. *Carbohydr. Polym.* **2016**, *151*, 9–19. [CrossRef] [PubMed]
19. Beigzadeh Ghelejlu, S.; Esmaiili, M.; Almasi, H. Characterization of chitosan-nanoclay bionanocomposite active films containing milk thistle extract. *Int. J. Biol. Macromol.* **2016**, *86*, 613–621. [CrossRef] [PubMed]
20. Wang, Y.C.; Lu, L.; Gunasekaran, S. Biopolymer/gold nanoparticles composite plasmonic thermal history indicator to monitor quality and safety of perishable bioproducts. *Biosens. Bioelectron.* **2017**, *92*, 109–116. [CrossRef] [PubMed]
21. Suh, S.; Meng, X.; Ko, S. Proof of concept study for different-sized chitosan nanoparticles as carbon dioxide (CO_2) indicators in food quality monitoring. *Talanta* **2016**, *161*, 265–270. [CrossRef] [PubMed]
22. Wyrwa, J.; Barska, A. Innovations in the food packaging market: Active packaging. *Eur. Food Res. Technol.* **2017**, *243*, 1681–1692. [CrossRef]
23. Cwiek-Ludwicka, K.; Ludwicki, J.K. Nanomaterials in food contact materials; considerations for risk assessment. *Rocz. Państwowego Zakładu Hig.* **2017**, *68*, 321–329.
24. Piperigkou, Z.; Karamanou, K.; Engin, A.B.; Gialeli, C.; Docea, A.O.; Vynios, D.H.; Pavao, M.S.G.; Golokhvast, K.S.; Shtilman, M.I.; Argiris, A.; et al. Emerging aspects of nanotoxicology in health and disease: From agriculture and food sector to cancer therapeutics. *Food Chem. Toxicol.* **2016**, *91*, 42–57. [CrossRef] [PubMed]
25. Hoseinnejad, M.; Jafari, S.M.; Katouzian, I. Inorganic and metal nanoparticles and their antimicrobial activity in food packaging applications. *Crit. Rev. Microbiol.* **2018**, *44*, 161–181. [CrossRef] [PubMed]
26. Almasi, H.; Jafarzadeh, P.; Mehryar, L. Fabrication of novel nanohybrids by impregnation of CuO nanoparticles into bacterial cellulose and chitosan nanofibers: Characterization, antimicrobial and release properties. *Carbohydr. Polym.* **2018**, *186*, 273–281. [CrossRef] [PubMed]
27. Attaran, S.A.; Hassan, A.; Wahit, M.U. Materials for food packaging applications based on bio-based polymer nanocomposites: A review. *J. Thermoplast. Compos. Mater.* **2017**, *30*, 143–173. [CrossRef]
28. Perinelli, D.R.; Fagioli, L.; Campana, R.; Lam, J.K.W.; Baffone, W.; Palmieri, G.F.; Casettari, L.; Bonacucina, G. Chitosan-based nanosystems and their exploited antimicrobial activity. *Eur. J. Pharm. Sci.* **2018**, *117*, 8–20. [CrossRef] [PubMed]
29. Torres-Giner, S.; Wilkanowicz, S.; Melendez-Rodriguez, B.; Lagaron, J.M. Nanoencapsulation of Aloe vera in synthetic and naturally occurring polymers by electrohydrodynamic processing of interest in food technology and bioactive packaging. *J. Agric. Food Chem.* **2017**, *65*, 4439–4448. [CrossRef] [PubMed]
30. Mei, L.; Teng, Z.; Zhu, G.Z.; Liu, Y.J.; Zhang, F.W.; Zhang, J.L.; Li, Y.; Guan, Y.G.; Luo, Y.G.; Chen, X.G.; et al. Silver nanocluster-embedded zein films as antimicrobial coating materials for food packaging. *ACS Appl. Mater. Interfaces* **2017**, *9*, 35297–35304. [CrossRef] [PubMed]
31. Shankar, S.; Wang, L.F.; Rhim, J.W. Preparations and characterization of alginate/silver composite films: Effect of types of silver particles. *Carbohydr. Polym.* **2016**, *146*, 208–216. [CrossRef] [PubMed]
32. Zhang, X.F.; Liu, Z.G.; Shen, W.; Gurunathan, S. Silver nanoparticles: Synthesis, characterization, properties, applications, and therapeutic approaches. *Int. J. Mol. Sci.* **2016**, *17*, 1534. [CrossRef] [PubMed]

33. Chu, Z.; Zhao, T.; Li, L.; Fan, J.; Qin, Y. Characterization of antimicrobial poly(lactic acid)/nano-composite films with silver and zinc oxide nanoparticles. *Materials* **2017**, *10*, 659. [CrossRef] [PubMed]

34. Tao, G.; Cai, R.; Wang, Y.J.; Song, K.; Guo, P.C.; Zhao, P.; Zuo, H.; He, H.W. Biosynthesis and characterization of AgNPs-silk/PVA film for potential packaging application. *Materials* **2017**, *10*, 667. [CrossRef] [PubMed]

35. Shao, Y.; Wu, C.; Wu, T.; Yuan, C.; Chen, S.; Ding, T.; Ye, X.; Hu, Y. Green synthesis of sodium alginate-silver nanoparticles and their antibacterial activity. *Int. J. Biol. Macromol.* **2018**, *111*, 1281–1292. [CrossRef] [PubMed]

36. Narayanan, K.B.; Han, S.S. Dual-crosslinked poly(vinyl alcohol)/sodium alginate/silver nanocomposite beads—A promising antimicrobial material. *Food Chem.* **2017**, *234*, 103–110. [CrossRef] [PubMed]

37. Patra, J.K.; Das, G.; Baek, K.H. Phyto-mediated biosynthesis of silver nanoparticles using the rind extract of watermelon (citrullus lanatus) under photo-catalyzed condition and investigation of its antibacterial, anticandidal and antioxidant efficacy. *J. Photochem. Photobiol. B* **2016**, *161*, 200–210. [CrossRef] [PubMed]

38. Azlin-Hasim, S.; Cruz-Romero, M.C.; Cummins, E.; Kerry, J.P.; Morris, M.A. The potential use of a layer-by-layer strategy to develop ldpe antimicrobial films coated with silver nanoparticles for packaging applications. *J. Colloid Interface Sci.* **2016**, *461*, 239–248. [CrossRef] [PubMed]

39. Arfat, Y.A.; Ahmed, J.; Hiremath, N.; Auras, R.; Joseph, A. Thermo-mechanical, rheological, structural and antimicrobial properties of bionanocomposite films based on fish skin gelatin and silver-copper nanoparticles. *Food Hydrocoll.* **2017**, *62*, 191–202. [CrossRef]

40. Jafari, H.; Pirouzifard, M.; Khaledabad, M.A.; Almasi, H. Effect of chitin nanofiber on the morphological and physical properties of chitosan/silver nanoparticle bionanocomposite films. *Int. J. Biol. Macromol.* **2016**, *92*, 461–466. [CrossRef] [PubMed]

41. Ramachandraiah, K.; Gnoc, N.T.B.; Chin, K.B. Biosynthesis of silver nanoparticles from persimmon byproducts and incorporation in biodegradable sodium alginate thin film. *J. Food Sci.* **2017**, *82*, 2329–2336. [CrossRef] [PubMed]

42. Heli, B.; Morales-Narvaez, E.; Golmohammadi, H.; Ajji, A.; Merkoci, A. Modulation of population density and size of silver nanoparticles embedded in bacterial cellulose via ammonia exposure: Visual detection of volatile compounds in a piece of plasmonic nanopaper. *Nanoscale* **2016**, *8*, 7984–7991. [CrossRef] [PubMed]

43. Tavakoli, H.; Rastegar, H.; Taherian, M.; Samadi, M.; Rostami, H. The effect of nano-silver packaging in increasing the shelf life of nuts: An in vitro model. *Ital. J. Food Saf.* **2017**, *6*, 6874. [CrossRef] [PubMed]

44. Deus, D.; Kehrenberg, C.; Schaudien, D.; Klein, G.; Krischek, C. Effect of a nano-silver coating on the quality of fresh turkey meat during storage after modified atmosphere or vacuum packaging. *Poult. Sci.* **2017**, *96*, 449–457. [CrossRef] [PubMed]

45. Ahmed, J.; Arfat, Y.A.; Bher, A.; Mulla, M.; Jacob, H.; Auras, R. Active chicken meat packaging based on polylactide films and bimetallic Ag-Cu nanoparticles and essential oil. *J. Food Sci.* **2018**, *83*, 1299–1310. [CrossRef] [PubMed]

46. Stormer, A.; Bott, J.; Kemmer, D.; Franz, R. Critical review of the migration potential of nanoparticles in food contact plastics. *Trends Food Sci. Technol.* **2017**, *63*, 39–50. [CrossRef]

47. Gallocchio, F.; Cibin, V.; Biancotto, G.; Roccato, A.; Muzzolon, O.; Carmen, L.; Simone, B.; Manodori, L.; Fabrizi, A.; Patuzzi, I.; et al. Testing nano-silver food packaging to evaluate silver migration and food spoilage bacteria on chicken meat. *Food Addit. Contam. A* **2016**, *33*, 1063–1071. [CrossRef] [PubMed]

48. Tiimob, B.J.; Mwinyelle, G.; Abdela, W.; Samuel, T.; Jeelani, S.; Rangari, V.K. Nanoengineered eggshell-silver tailored copolyester polymer blend film with antimicrobial properties. *J. Agric. Food Chem.* **2017**, *65*, 1967–1976. [CrossRef] [PubMed]

49. Su, Q.Z.; Lin, Q.B.; Chen, C.F.; Wu, L.B.; Wang, Z.W. Effect of organic additives on silver release from nanosilver-polyethylene composite films to acidic food simulant. *Food Chem.* **2017**, *228*, 560–566. [CrossRef] [PubMed]

50. Hosseini, R.; Ahari, H.; Mahasti, P.; Paidari, S. Measuring the migration of silver from silver nanocomposite polyethylene packaging based on (TiO$_2$) into penaeus semisulcatus using titration comparison with migration methods. *Fish. Sci.* **2017**, *83*, 649–659. [CrossRef]

51. Hannon, J.C.; Kerry, J.P.; Cruz-Romero, M.; Azlin-Hasim, S.; Morris, M.; Cummins, E. Migration assessment of silver from nanosilver spray coated low density polyethylene or polyester films into milk. *Food Packag. Shelf Life* **2018**, *15*, 144–150. [CrossRef]

52. Becaro, A.A.; Siqueira, M.C.; Puti, F.C.; de Moura, M.R.; Correa, D.S.; Marconcini, J.M.; Mattoso, L.H.C.; Ferreira, M.D. Cytotoxic and genotoxic effects of silver nanoparticle/carboxymethyl cellulose on allium cepa. *Environ. Monit. Assess.* **2017**, *189*. [CrossRef] [PubMed]

53. Mikiciuk, J.; Mikiciuk, E.; Wronska, A.; Szterk, A. Antimicrobial potential of commercial silver nanoparticles and the characterization of their physical properties toward probiotic bacteria isolated from fermented milk products. *J. Environ. Sci. Health B* **2016**, *51*, 222–229. [CrossRef] [PubMed]

54. Krol, A.; Pomastowski, P.; Rafinska, K.; Railean-Plugaru, V.; Buszewski, B. Zinc oxide nanoparticles: Synthesis, antiseptic activity and toxicity mechanism. *Adv. Colloid Interface Sci.* **2017**, *249*, 37–52. [CrossRef] [PubMed]

55. Rokbani, H.; Daigle, F.; Ajji, A. Combined effect of ultrasound stimulations and autoclaving on the enhancement of antibacterial activity of ZnO and SiO_2/ZnO nanoparticles. *Nanomaterials* **2018**, *8*, 129. [CrossRef] [PubMed]

56. Jafarzadeh, S.; Ariffin, F.; Mahmud, S.; Alias, A.; Hosseini, S.F.; Ahmad, M. Improving the physical and protective functions of semolina films by embedding a blend nanofillers (ZnO-nr and nano-kaolin). *Food Packag. Shelf Life* **2017**, *12*, 66–75. [CrossRef]

57. Salarbashi, D.; Mortazavi, S.A.; Noghabi, M.S.; Fazly Bazzaz, B.S.; Sedaghat, N.; Ramezani, M.; Shahabi-Ghahfarrokhi, I. Development of new active packaging film made from a soluble soybean polysaccharide incorporating ZnO nanoparticles. *Carbohydr. Polym.* **2016**, *140*, 220–227. [CrossRef] [PubMed]

58. Shahmohammadi Jebel, F.; Almasi, H. Morphological, physical, antimicrobial and release properties of ZnO nanoparticles-loaded bacterial cellulose films. *Carbohydr. Polym.* **2016**, *149*, 8–19. [CrossRef] [PubMed]

59. Akbariazam, M.; Ahmadi, M.; Javadian, N.; Mohammadi Nafchi, A. Fabrication and characterization of soluble soybean polysaccharide and nanorod-rich ZnO bionanocomposite. *Int. J. Biol. Macromol.* **2016**, *89*, 369–375. [CrossRef] [PubMed]

60. Esmailzadeh, H.; Sangpour, P.; Shahraz, F.; Hejazi, J.; Khaksar, R. Effect of nanocomposite packaging containing ZnO on growth of bacillus subtilis and enterobacter aerogenes. *Mater. Sci. Eng. C Mater. Biol. Appl.* **2016**, *58*, 1058–1063. [CrossRef] [PubMed]

61. Mizielinska, M.; Kowalska, U.; Jarosz, M.; Suminska, P.; Landercy, N.; Duquesne, E. The effect of UV aging on antimicrobial and mechanical properties of pla films with incorporated zinc oxide nanoparticles. *Int. J. Environ. Res. Public Health* **2018**, *15*, 794. [CrossRef] [PubMed]

62. Kotharangannagari, V.K.; Krishnan, K. Biodegradable hybrid nanocomposites of starch/lysine and ZnO nanoparticles with shape memory properties. *Mater. Des.* **2016**, *109*, 590–595. [CrossRef]

63. Babaei-Ghazvini, A.; Shahabi-Ghahfarrokhi, I.; Goudarzi, V. Preparation of UV-protective starch/kefiran/ZnO nanocomposite as a packaging film: Characterization. *Food Packag. Shelf Life* **2018**, *16*, 103–111. [CrossRef]

64. Mizielinska, M.; Kowalska, U.; Jarosz, M.; Suminska, P. A comparison of the effects of packaging containing nano ZnO or polylysine on the microbial purity and texture of Cod (gadus morhua) fillets. *Nanomaterials* **2018**, *8*, 158. [CrossRef] [PubMed]

65. Calderon, V.S.; Gomes, B.; Ferreira, P.J.; Carvalho, S. Zinc nanostructures for oxygen scavenging. *Nanoscale* **2017**, *9*, 5254–5262. [CrossRef] [PubMed]

66. Li, W.; Li, L.; Cao, Y.; Lan, T.; Chen, H.; Qin, Y. Effects of pla film incorporated with ZnO nanoparticle on the quality attributes of fresh-cut apple. *Nanomaterials* **2017**, *7*, 207. [CrossRef] [PubMed]

67. Beak, S.; Kim, H.; Song, K.B. Characterization of an olive flounder bone gelatin-zinc oxide nanocomposite film and evaluation of its potential application in spinach packaging. *J. Food Sci.* **2017**, *82*, 2643–2649. [CrossRef] [PubMed]

68. Suo, B.; Li, H.; Wang, Y.; Li, Z.; Pan, Z.; Ai, Z. Effects of ZnO nanoparticle-coated packaging film on pork meat quality during cold storage. *J. Sci. Food Agric.* **2017**, *97*, 2023–2029. [CrossRef] [PubMed]

69. Al-Shabib, N.A.; Husain, F.M.; Ahmed, F.; Khan, R.A.; Ahmad, I.; Alsharaeh, E.; Khan, M.S.; Hussain, A.; Rehman, M.T.; Yusuf, M.; et al. Biogenic synthesis of zinc oxide nanostructures from nigella sativa seed: Prospective role as food packaging material inhibiting broad-spectrum quorum sensing and biofilm. *Sci. Rep.* **2016**, *6*, 36761. [CrossRef] [PubMed]

70. Ansar, S.; Abudawood, M.; Hamed, S.S.; Aleem, M.M. Exposure to zinc oxide nanoparticles induces neurotoxicity and proinflammatory response: Amelioration by hesperidin. *Biol. Trace Elem. Res.* **2017**, *175*, 360–366. [CrossRef] [PubMed]

71.	Senapati, V.A.; Gupta, G.S.; Pandey, A.K.; Shanker, R.; Dhawan, A.; Kumar, A. Zinc oxide nanoparticle induced age dependent immunotoxicity in BALB/c mice. *Toxicol. Res.-Uk* **2017**, *6*, 342–352. [CrossRef] [PubMed]

72.	Moreno-Olivas, F.; Tako, E.; Mahler, G.J. Zno nanoparticles affect intestinal function in an in vitro model. *Food Funct.* **2018**, *9*, 1475–1491. [CrossRef] [PubMed]

73.	Zhang, H.; Bussini, D.; Hortal, M.; Elegir, G.; Mendes, J.; Jorda Beneyto, M. PLA coated paper containing active inorganic nanoparticles: Material characterization and fate of nanoparticles in the paper recycling process. *Waste Manag.* **2016**, *52*, 339–345. [CrossRef] [PubMed]

74.	Chia, S.L.; Leong, D.T. Reducing ZnO nanoparticles toxicity through silica coating. *Heliyon* **2016**, *2*, e00177. [CrossRef] [PubMed]

75.	Grigore, M.E.; Biscu, E.R.; Holban, A.M.; Gestal, M.C.; Grumezescu, A.M. Methods of synthesis, properties and biomedical applications of CuO nanoparticles. *Pharmaceuticals* **2016**, *9*, 75. [CrossRef] [PubMed]

76.	Gu, H.D.; Chen, X.; Chen, F.; Zhou, X.; Parsaee, Z. Ultrasound-assisted biosynthesis of CuO-NPs using brown alga cystoseira trinodis: Characterization, photocatalytic AOP, DPPH scavenging and antibacterial investigations. *Ultrason. Sonochem.* **2018**, *41*, 109–119. [CrossRef] [PubMed]

77.	Eivazihollagh, A.; Backstrom, J.; Dahlstrom, C.; Carlsson, F.; Ibrahem, I.; Lindman, B.; Edlund, H.; Norgren, M. One-pot synthesis of cellulose-templated copper nanoparticles with antibacterial properties. *Mater. Lett.* **2017**, *187*, 170–172. [CrossRef]

78.	Castro Mayorga, J.L.; Fabra Rovira, M.J.; Cabedo Mas, L.; Sanchez Moragas, G.; Lagaron Cabello, J.M. Antimicrobial nanocomposites and electrospun coatings based on poly(3-hydroxybutyrate-*co*-3-hydroxyvalerate) and copper oxide nanoparticles for active packaging and coating applications. *J. Appl. Polym. Sci.* **2018**, *135*. [CrossRef]

79.	Gautam, G.; Mishra, P. Development and characterization of copper nanocomposite containing bilayer film for coconut oil packaging. *J. Food Process. Preserv.* **2017**, *41*, e13243. [CrossRef]

80.	Beigmohammadi, F.; Peighambardoust, S.H.; Hesari, J.; Azadmard-Damirchi, S.; Peighambardoust, S.J.; Khosrowshahi, N.K. Antibacterial properties of LDPE nanocomposite films in packaging of UF cheese. *LWT-Food Sci. Technol.* **2016**, *65*, 106–111. [CrossRef]

81.	Shankar, S.; Wang, L.F.; Rhim, J.W. Preparation and properties of carbohydrate-based composite films incorporated with CuO nanoparticles. *Carbohydr. Polym.* **2017**, *169*, 264–271. [CrossRef] [PubMed]

82.	Li, K.; Jin, S.C.; Liu, X.R.; Chen, H.; He, J.; Li, J.Z. Preparation and characterization of chitosan/soy protein isolate nanocomposite film reinforced by Cu nanoclusters. *Polymers* **2017**, *9*, 247. [CrossRef]

83.	Lomate, G.B.; Dandi, B.; Mishra, S. Development of antimicrobial LDPE/Cu nanocomposite food packaging film for extended shelf life of peda. *Food Packag. Shelf Life* **2018**, *16*, 211–219. [CrossRef]

84.	Tamayo, L.; Azocar, M.; Kogan, M.; Riveros, A.; Paez, M. Copper-polymer nanocomposites: An excellent and cost-effective biocide for use on antibacterial surfaces. *Mater. Sci. Eng. C Mater. Biol. Appl.* **2016**, *69*, 1391–1409. [CrossRef] [PubMed]

85.	Yadav, H.M.; Kim, J.S.; Pawar, S.H. Developments in photocatalytic antibacterial activity of nano TiO_2: A review. *Korean J. Chem. Eng.* **2016**, *33*, 1989–1998. [CrossRef]

86.	Zhang, W.; Chen, J.; Chen, Y.; Xia, W.; Xiong, Y.L.; Wang, H. Enhanced physicochemical properties of chitosan/whey protein isolate composite film by sodium laurate-modified TiO_2 nanoparticles. *Carbohydr. Polym.* **2016**, *138*, 59–65. [CrossRef] [PubMed]

87.	He, Q.; Zhang, Y.; Cai, X.; Wang, S. Fabrication of gelatin-TiO_2 nanocomposite film and its structural, antibacterial and physical properties. *Int. J. Biol. Macromol.* **2016**, *84*, 153–160. [CrossRef] [PubMed]

88.	Li, H.; Yang, J.; Li, P.; Lan, T.; Peng, L. A facile method for preparation superhydrophobic paper with enhanced physical strength and moisture-proofing property. *Carbohydr. Polym.* **2017**, *160*, 9–17. [CrossRef] [PubMed]

89.	Lopez de Dicastillo, C.; Patino, C.; Galotto, M.J.; Palma, J.L.; Alburquenque, D.; Escrig, J. Novel antimicrobial titanium dioxide nanotubes obtained through a combination of atomic layer deposition and electrospinning technologies. *Nanomaterials* **2018**, *8*, 128. [CrossRef] [PubMed]

90.	Nesic, A.; Gordic, M.; Davidovic, S.; Radovanovic, Z.; Nedeljkovic, J.; Smirnova, I.; Gurikov, P. Pectin-based nanocomposite aerogels for potential insulated food packaging application. *Carbohydr. Polym.* **2018**, *195*, 128–135. [CrossRef] [PubMed]

91. Xing, Y.G.; Li, X.H.; Zhang, L.; Xu, Q.L.; Che, Z.M.; Li, W.L.; Bai, Y.M.; Li, K. Effect of TiO_2 nanoparticles on the antibacterial and physical properties of polyethylene-based film. *Prog. Org. Coat.* **2012**, *73*, 219–224. [CrossRef]

92. Roilo, D.; Maestri, C.A.; Scarpa, M.; Bettotti, P.; Checchetto, R. Gas barrier and optical properties of cellulose nanofiber coatings with dispersed TiO_2 nanoparticles. *Surf. Coat. Technol.* **2018**, *343*, 131–137. [CrossRef]

93. Oleyaei, S.A.; Zahedi, Y.; Ghanbarzadeh, B.; Moayedi, A.A. Modification of physicochemical and thermal properties of starch films by incorporation of TiO_2 nanoparticles. *Int. J. Biol. Macromol.* **2016**, *89*, 256–264. [CrossRef] [PubMed]

94. Goudarzi, V.; Shahabi-Ghahfarrokhi, I.; Babaei-Ghazvini, A. Preparation of ecofriendly UV-protective food packaging material by starch/TiO_2 bio-nanocomposite: Characterization. *Int. J. Biol. Macromol.* **2017**, *95*, 306–313. [CrossRef] [PubMed]

95. Abdel Rehim, M.H.; El-Samahy, M.A.; Badawy, A.A.; Mohram, M.E. Photocatalytic activity and antimicrobial properties of paper sheets modified with TiO_2/sodium alginate nanocomposites. *Carbohydr. Polym.* **2016**, *148*, 194–199. [CrossRef] [PubMed]

96. Mihaly-Cozmuta, A.; Peter, A.; Craciun, G.; Falup, A.; Mihaly-Cozmuta, L.; Nicula, C.; Vulpoi, A.; Baia, M. Preparation and characterization of active cellulose-based papers modified with TiO_2, Ag and zeolite nanocomposites for bread packaging application. *Cellulose* **2017**, *24*, 3911–3928. [CrossRef]

97. Li, D.; Ye, Q.; Jiang, L.; Luo, Z. Effects of nano-TiO_2-LDPE packaging on postharvest quality and antioxidant capacity of strawberry (*Fragaria ananassa* Duch.) stored at refrigeration temperature. *J. Sci. Food Agric.* **2017**, *97*, 1116–1123. [CrossRef] [PubMed]

98. Winkler, H.C.; Notter, T.; Meyer, U.; Naegeli, H. Critical review of the safety assessment of titanium dioxide additives in food. *J. Nanobiotechnol.* **2018**, *16*, 51. [CrossRef] [PubMed]

99. Ozgur, M.E.; Balcioglu, S.; Ulu, A.; Ozcan, I.; Okumus, F.; Koytepe, S.; Ates, B. The in vitro toxicity analysis of titanium dioxide (TiO_2) nanoparticles on kinematics and biochemical quality of rainbow trout sperm cells. *Environ. Toxicol. Pharmacol.* **2018**, *62*, 11–19. [CrossRef] [PubMed]

100. Salarbashi, D.; Tafaghodi, M.; Bazzaz, B.S.F. Soluble soybean polysaccharide/TiO_2 bionanocomposite film for food application. *Carbohydr. Polym.* **2018**, *186*, 384–393. [CrossRef] [PubMed]

101. Jo, M.R.; Yu, J.; Kim, H.J.; Song, J.H.; Kim, K.M.; Oh, J.M.; Choi, S.J. Titanium dioxide nanoparticle-biomolecule interactions influence oral absorption. *Nanomaterials* **2016**, *6*, 225. [CrossRef] [PubMed]

102. Swaroop, C.; Shukla, M. Nano-magnesium oxide reinforced polylactic acid biofilms for food packaging applications. *Int. J. Biol. Macromol.* **2018**, *113*, 729–736. [CrossRef] [PubMed]

103. Shariatinia, Z.; Fazli, M. Mechanical properties and antibacterial activities of novel nanobiocomposite films of chitosan and starch. *Food Hydrocoll.* **2015**, *46*, 112–124. [CrossRef]

104. Ciabocco, M.; Cancemi, P.; Saladino, M.L.; Caponetti, E.; Alduina, R.; Berrettoni, M. Synthesis and antibacterial activity of iron-hexacyanocobaltate nanoparticles. *J. Biol. Inorg. Chem.* **2018**, *23*, 385–398. [CrossRef] [PubMed]

105. Liu, S.; Li, X.; Chen, L.; Li, L.; Li, B.; Zhu, J.; Liang, X. Investigating the H_2O/O_2 selective permeability from a view of multi-scale structure of starch/SiO_2 nanocomposites. *Carbohydr. Polym.* **2017**, *173*, 143–149. [CrossRef] [PubMed]

106. Ren, P.-G.; Wang, H.; Yan, D.-X.; Huang, H.-D.; Wang, H.-B.; Zhang, Z.-P.; Xu, L.; Li, Z.-M. Ultrahigh gas barrier poly(vinyl alcohol) nanocomposite film filled with congregated and oriented Fe_3O_4@GO sheets induced by magnetic-field. *Compos. Part A-Appl. Sci. Manuf.* **2017**, *97*, 1–9. [CrossRef]

107. Khalaj, M.J.; Ahmadi, H.; Lesankhosh, R.; Khalaj, G. Study of physical and mechanical properties of polypropylene nanocomposites for food packaging application: Nano-clay modified with iron nanoparticles. *Trends Food Sci. Technol.* **2016**, *51*, 41–48. [CrossRef]

108. Mallakpour, S.; Nazari, H.Y. The influence of bovine serum albumin-modified silica on the physicochemical properties of poly(vinyl alcohol) nanocomposites synthesized by ultrasonication technique. *Ultrason. Sonochem.* **2018**, *41*, 1–10. [CrossRef] [PubMed]

109. Guo, Z.; Martucci, N.J.; Liu, Y.; Yoo, E.; Tako, E.; Mahler, G.J. Silicon dioxide nanoparticle exposure affects small intestine function in an in vitro model. *Nanotoxicology* **2018**, *12*, 485–508. [CrossRef] [PubMed]

110. Uddin, F. Clays, nanoclays, and montmorillonite minerals. *Metall. Mater. Trans. A* **2008**, *39*, 2804–2814. [CrossRef]

111. Gutiérrez, T.J.; Ponce, A.G.; Alvarez, V.A. Nano-clays from natural and modified montmorillonite with and without added blueberry extract for active and intelligent food nanopackaging materials. *Mater. Chem. Phys.* **2017**, *194*, 283–292. [CrossRef]

112. Jang, S.H.; Jang, S.R.; Lee, G.M.; Ryu, J.H.; Park, S.I.; Park, N.H. Halloysite nanocapsules containing thyme essential oil: Preparation, characterization, and application in packaging materials. *J. Food Sci.* **2017**, *82*, 2113–2120. [CrossRef] [PubMed]

113. Pereira, R.C.; Carneiro, J.D.S.; Assis, O.B.; Borges, S.V. Mechanical and structural characterization of whey protein concentrate/montmorillonite/lycopene films. *J. Sci. Food Agric.* **2017**, *97*, 4978–4986. [CrossRef] [PubMed]

114. Orsuwan, A.; Sothornvit, R. Development and characterization of banana flour film incorporated with montmorillonite and banana starch nanoparticles. *Carbohydr. Polym.* **2017**, *174*, 235–242. [CrossRef] [PubMed]

115. Oliveira, T.I.; Zea-Redondo, L.; Moates, G.K.; Wellner, N.; Cross, K.; Waldron, K.W.; Azeredo, H.M. Pomegranate peel pectin films as affected by montmorillonite. *Food Chem.* **2016**, *198*, 107–112. [CrossRef] [PubMed]

116. Zahedi, Y.; Fathi-Achachlouei, B.; Yousefi, A.R. Physical and mechanical properties of hybrid montmorillonite/zinc oxide reinforced carboxymethyl cellulose nanocomposites. *Int. J. Biol. Macromol.* **2018**, *108*, 863–873. [CrossRef] [PubMed]

117. Kim, J.M.; Lee, M.H.; Ko, J.A.; Kang, D.H.; Bae, H.; Park, H.J. Influence of food with high moisture content on oxygen barrier property of polyvinyl alcohol (PVA)/vermiculite nanocomposite coated multilayer packaging film. *J. Food Sci.* **2018**, *83*, 349–357. [CrossRef] [PubMed]

118. Lee, M.H.; Seo, H.S.; Park, H.J. Thyme oil encapsulated in halloysite nanotubes for antimicrobial packaging system. *J. Food Sci.* **2017**, *82*, 922–932. [CrossRef] [PubMed]

119. Peter, A.; Mihaly-Cozmuta, L.; Mihaly-Cozmuta, A.; Nicula, C.; Ziemkowska, W.; Basiak, D.; Danciu, V.; Vulpoi, A.; Baia, L.; Falup, A.; et al. Changes in the microbiological and chemical characteristics of white bread during storage in paper packages modified with Ag/TiO$_2$-SiO$_2$, Ag/N-TiO$_2$ or Au/TiO$_2$. *Food Chem.* **2016**, *197*, 790–798. [CrossRef] [PubMed]

120. Nalcabasmaz, S.; Ayhan, Z.; Cimmino, S.; Silvestre, C.; Duraccio, D. Effects of pp-based nanopackaging on the overall quality and shelf life of ready-to-eat salami. *Packag. Technol. Sci.* **2017**, *30*, 663–679. [CrossRef]

121. Kim, J.; Park, N.H.; Na, J.H.; Han, J. Development of natural insect-repellent loaded halloysite nanotubes and their application to food packaging to prevent plodia interpunctella infestation. *J. Food Sci.* **2016**, *81*, E1956–E1965. [CrossRef] [PubMed]

122. Peighambardoust, S.H.; Beigmohammadi, F.; Peighambardoust, S.J. Application of organoclay nanoparticle in low-density polyethylene films for packaging of UF cheese. *Packag. Technol. Sci.* **2016**, *29*, 355–363. [CrossRef]

123. Echeverria, I.; Lopez-Caballero, M.E.; Gomez-Guillen, M.C.; Mauri, A.N.; Montero, M.P. Active nanocomposite films based on soy proteins-montmorillonite-clove essential oil for the preservation of refrigerated bluefin tuna (thunnus thynnus) fillets. *Int. J. Food Microbiol.* **2018**, *266*, 142–149. [CrossRef] [PubMed]

124. Guimaraes, I.C.; dos Reis, K.C.; Menezes, E.G.; Borges, P.R.; Rodrigues, A.C.; Leal, R.; Hernandes, T.; de Carvalho, E.H.; Vilas Boas, E.V. Combined effect of starch/montmorillonite coating and passive map in antioxidant activity, total phenolics, organic acids and volatile of fresh-cut carrots. *Int. J. Food Sci. Nutr.* **2016**, *67*, 141–152. [CrossRef] [PubMed]

125. Junqueira-Goncalves, M.P.; Salinas, G.E.; Bruna, J.E.; Niranjan, K. An assessment of lactobiopolymer-montmorillonite composites for dip coating applications on fresh strawberries. *J. Sci. Food Agric.* **2017**, *97*, 1846–1853. [CrossRef] [PubMed]

126. Wagner, A.; Eldawud, R.; White, A.; Agarwal, S.; Stueckle, T.A.; Sierros, K.A.; Rojanasakul, Y.; Gupta, R.K.; Dinu, C.Z. Toxicity evaluations of nanoclays and thermally degraded byproducts through spectroscopical and microscopical approaches. *Biochim. Biophys. Acta* **2017**, *1861*, 3406–3415. [CrossRef] [PubMed]

127. Han, C.; Zhao, A.; Varughese, E.; Sahle-Demessie, E. Evaluating weathering of food packaging polyethylene-nano-clay composites: Release of nanoparticles and their impacts. *NanoImpact* **2018**, *9*, 61–71. [CrossRef] [PubMed]

128. Echegoyen, Y.; Rodriguez, S.; Nerin, C. Nanoclay migration from food packaging materials. *Food Addit. Contam. Part A Chem. Anal. Control Expo. Risk Assess.* **2016**, *33*, 530–539. [CrossRef] [PubMed]

129. Lambert, S.; Wagner, M. Environmental performance of bio-based and biodegradable plastics: The road ahead. *Chem. Soc. Rev.* **2017**, *46*, 6855–6871. [CrossRef] [PubMed]

130. Yang, W.; Fortunati, E.; Bertoglio, F.; Owczarek, J.S.; Bruni, G.; Kozanecki, M.; Kenny, J.M.; Torre, L.; Visai, L.; Puglia, D. Polyvinyl alcohol/chitosan hydrogels with enhanced antioxidant and antibacterial properties induced by lignin nanoparticles. *Carbohydr. Polym.* **2018**, *181*, 275–284. [CrossRef] [PubMed]

131. Sarwar, M.S.; Niazi, M.B.K.; Jahan, Z.; Ahmad, T.; Hussain, A. Preparation and characterization of PVA/nanocellulose/Ag nanocomposite films for antimicrobial food packaging. *Carbohydr. Polym.* **2018**, *184*, 453–464. [CrossRef] [PubMed]

132. Rouhi, M.; Razavi, S.H.; Mousavi, S.M. Optimization of crosslinked poly(vinyl alcohol) nanocomposite films for mechanical properties. *Mater. Sci. Eng. C Mater. Biol. Appl.* **2017**, *71*, 1052–1063. [CrossRef] [PubMed]

133. El Achaby, M.; El Miri, N.; Aboulkas, A.; Zahouily, M.; Bilal, E.; Barakat, A.; Solhy, A. Processing and properties of eco-friendly bio-nanocomposite films filled with cellulose nanocrystals from sugarcane bagasse. *Int. J. Biol. Macromol.* **2017**, *96*, 340–352. [CrossRef] [PubMed]

134. Giannakas, A.; Vlacha, M.; Salmas, C.; Leontiou, A.; Katapodis, P.; Stamatis, H.; Barkoula, N.M.; Ladavos, A. Preparation, characterization, mechanical, barrier and antimicrobial properties of chitosan/PVOH/clay nanocomposites. *Carbohydr. Polym.* **2016**, *140*, 408–415. [CrossRef] [PubMed]

135. Rezaeigolestani, M.; Misaghi, A.; Khanjari, A.; Basti, A.A.; Abdulkhani, A.; Fayazfar, S. Antimicrobial evaluation of novel poly-lactic acid based nanocomposites incorporated with bioactive compounds in-vitro and in refrigerated vacuum-packed cooked sausages. *Int. J. Food Microbiol.* **2017**, *260*, 1–10. [CrossRef] [PubMed]

136. Vasile, C.; Rapa, M.; Stefan, M.; Stan, M.; Macavei, S.; Darie-Nita, R.N.; Barbu-Tudoran, L.; Vodnar, D.C.; Popa, E.E.; Stefan, R.; et al. New PLA/ZnO:Cu/Ag bionanocomposites for food packaging. *Express Polym. Lett.* **2017**, *11*, 531–544. [CrossRef]

137. Aframehr, W.M.; Molki, B.; Heidarian, P.; Behzad, T.; Sadeghi, M.; Bagheri, R. Effect of calcium carbonate nanoparticles on barrier properties and biodegradability of polylactic acid. *Fibers Polym.* **2017**, *18*, 2041–2048. [CrossRef]

138. Shavisi, N.; Khanjari, A.; Basti, A.A.; Misaghi, A.; Shahbazi, Y. Effect of PLA films containing propolis ethanolic extract, cellulose nanoparticle and ziziphora clinopodioides essential oil on chemical, microbial and sensory properties of minced beef. *Meat Sci.* **2017**, *124*, 95–104. [CrossRef] [PubMed]

139. Wen, P.; Zhu, D.H.; Feng, K.; Liu, F.J.; Lou, W.Y.; Li, N.; Zong, M.H.; Wu, H. Fabrication of electrospun polylactic acid nanofilm incorporating cinnamon essential oil/β-cyclodextrin inclusion complex for antimicrobial packaging. *Food Chem.* **2016**, *196*, 996–1004. [CrossRef] [PubMed]

140. Castro-Mayorga, J.L.; Freitas, F.; Reis, M.A.M.; Prieto, M.A.; Lagaron, J.M. Biosynthesis of silver nanoparticles and polyhydroxybutyrate nanocomposites of interest in antimicrobial applications. *Int. J. Biol. Macromol.* **2018**, *108*, 426–435. [CrossRef] [PubMed]

141. Kuntzler, S.G.; Almeida, A.C.A.; Costa, J.A.V.; Morais, M.G. Polyhydroxybutyrate and phenolic compounds microalgae electrospun nanofibers: A novel nanomaterial with antibacterial activity. *Int. J. Biol. Macromol.* **2018**, *113*, 1008–1014. [CrossRef] [PubMed]

142. Shakil, O.; Masood, F.; Yasin, T. Characterization of physical and biodegradation properties of poly-3-hydroxybutyrate-*co*-3-hydroxyvalerate/sepiolite nanocomposites. *Mater. Sci. Eng. C Mater. Biol. Appl.* **2017**, *77*, 173–183. [CrossRef] [PubMed]

143. Gaaz, T.S.; Sulong, A.B.; Akhtar, M.N.; Kadhum, A.A.; Mohamad, A.B.; Al-Amiery, A.A. Properties and applications of polyvinyl alcohol, halloysite nanotubes and their nanocomposites. *Molecules* **2015**, *20*, 22833–22847. [CrossRef] [PubMed]

144. DeMerlis, C.C.; Schoneker, D.R. Review of the oral toxicity of polyvinyl alcohol (PVA). *Food Chem. Toxicol.* **2003**, *41*, 319–326. [CrossRef]

145. Sun, J.Y.; Shen, J.J.; Chen, S.K.; Cooper, M.A.; Fu, H.B.; Wu, D.M.; Yang, Z.G. Nanofiller reinforced biodegradable PLA/PHA composites: Current status and future trends. *Polymers* **2018**, *10*, 505. [CrossRef]

146. Zembouai, I.; Bruzaud, S.; Kaci, M.; Benhamida, A.; Corre, Y.M.; Grohens, Y.; Taguet, A.; Lopez-Cuesta, J.M. Poly(3-hydroxybutyrate-*co*-3-hydroxyvalerate)/polylactide blends: Thermal stability, flammability and thermo-mechanical behavior. *J. Polym. Environ.* **2014**, *22*, 131–139. [CrossRef]

147. Aqlil, M.; Moussemba Nzenguet, A.; Essamlali, Y.; Snik, A.; Larzek, M.; Zahouily, M. Graphene oxide filled lignin/starch polymer bionanocomposite: Structural, physical, and mechanical studies. *J. Agric. Food Chem.* **2017**, *65*, 10571–10581. [CrossRef] [PubMed]

148. Shahbazi, M.; Rajabzadeh, G.; Sotoodeh, S. Functional characteristics, wettability properties and cytotoxic effect of starch film incorporated with multi-walled and hydroxylated multi-walled carbon nanotubes. *Int. J. Biol. Macromol.* **2017**, *104*, 597–605. [CrossRef] [PubMed]

149. Shankar, S.; Rhim, J.W. Preparation of nanocellulose from micro-crystalline cellulose: The effect on the performance and properties of agar-based composite films. *Carbohydr. Polym.* **2016**, *135*, 18–26. [CrossRef] [PubMed]

150. Pal, N.; Dubey, P.; Gopinath, P.; Pal, K. Combined effect of cellulose nanocrystal and reduced graphene oxide into poly-lactic acid matrix nanocomposite as a scaffold and its anti-bacterial activity. *Int. J. Biol. Macromol.* **2017**, *95*, 94–105. [CrossRef] [PubMed]

151. Liu, S.; Li, X.; Chen, L.; Li, L.; Li, B.; Zhu, J. Tunable d-limonene permeability in starch-based nanocomposite films reinforced by cellulose nanocrystals. *J. Agric. Food Chem.* **2018**, *66*, 979–987. [CrossRef] [PubMed]

152. Lavoine, N.; Guillard, V.; Desloges, I.; Gontard, N.; Bras, J. Active bio-based food-packaging: Diffusion and release of active substances through and from cellulose nanofiber coating toward food-packaging design. *Carbohydr. Polym.* **2016**, *149*, 40–50. [CrossRef] [PubMed]

153. Postnova, I.; Silant'ev, V.; Sarin, S.; Shchipunov, Y. Chitosan hydrogels and bionanocomposites formed through the mineralization and regulated charging. *Chem. Rec.* **2018**. [CrossRef] [PubMed]

154. Liang, J.; Yan, H.; Zhang, J.; Dai, W.; Gao, X.; Zhou, Y.; Wan, X.; Puligundla, P. Preparation and characterization of antioxidant edible chitosan films incorporated with epigallocatechin gallate nanocapsules. *Carbohydr. Polym.* **2017**, *171*, 300–306. [CrossRef] [PubMed]

155. Buslovich, A.; Horev, B.; Rodov, V.; Gedanken, A.; Poverenov, E. One-step surface grafting of organic nanoparticles: In situ deposition of antimicrobial agents vanillin and chitosan on polyethylene packaging films. *J. Mater. Chem. B* **2017**, *5*, 2655–2661. [CrossRef]

156. Aytac, Z.; Ipek, S.; Durgun, E.; Tekinay, T.; Uyar, T. Antibacterial electrospun zein nanofibrous web encapsulating thymol/cyclodextrin-inclusion complex for food packaging. *Food Chem.* **2017**, *233*, 117–124. [CrossRef] [PubMed]

157. Rouf, T.B.; Schmidt, G.; Kokini, J.L. Zein-laponite nanocomposites with improved mechanical, thermal and barrier properties. *J. Mater. Sci.* **2018**, *53*, 7387–7402. [CrossRef]

158. Oymaci, P.; Altinkaya, S.A. Improvement of barrier and mechanical properties of whey protein isolate based food packaging films by incorporation of zein nanoparticles as a novel bionanocomposite. *Food Hydrocoll.* **2016**, *54*, 1–9. [CrossRef]

159. Gilbert, J.; Cheng, C.J.; Jones, O.G. Vapor barrier properties and mechanical behaviors of composite hydroxypropyl methylcelluose/zein nanoparticle films. *Food Biophys.* **2018**, *13*, 25–36. [CrossRef]

160. Qazanfarzadeh, Z.; Kadivar, M. Properties of whey protein isolate nanocomposite films reinforced with nanocellulose isolated from oat husk. *Int. J. Biol. Macromol.* **2016**, *91*, 1134–1140. [CrossRef] [PubMed]

161. Hassannia-Kolaee, M.; Khodaiyan, F.; Pourahmad, R.; Shahabi-Ghahfarrokhi, I. Development of ecofriendly bionanocomposite: Whey protein isolate/pullulan films with nano-SiO$_2$. *Int. J. Biol. Macromol.* **2016**, *86*, 139–144. [CrossRef] [PubMed]

162. Jamil, B.; Bokhari, H.; Imran, M. Mechanism of action: How nano-antimicrobials act? *Curr. Drug Targets* **2017**, *18*, 363–373. [CrossRef] [PubMed]

163. El Zowalaty, M.E.; Al Ali, S.H.H.; Husseiny, M.I.; Geilich, B.M.; Webster, T.J.; Hussein, M.Z. The ability of streptomycin-loaded chitosan-coated magnetic nanocomposites to possess antimicrobial and antituberculosis activities. *Int. J. Nanomed.* **2015**, *10*, 3269–3273. [CrossRef] [PubMed]

164. Dudefoi, W.; Villares, A.; Peyron, S.; Moreau, C.; Ropers, M.H.; Gontard, N.; Cathala, B. Nanoscience and nanotechnologies for biobased materials, packaging and food applications: New opportunities and concerns. *Innov. Food Sci. Emerg. Technol.* **2018**, *46*, 107–121. [CrossRef]

165. Souza, V.G.L.; Fernando, A.L. Nanoparticles in food packaging: Biodegradability and potential migration to food—A review. *Food Packag. Shelf Life* **2016**, *8*, 63–70. [CrossRef]

166. Jokar, M.; Pedersen, G.A.; Loeschner, K. Six open questions about the migration of engineered nano-objects from polymer-based food-contact materials: A review. *Food Addit. Contam. A* **2017**, *34*, 434–450. [CrossRef] [PubMed]

167. Ramos, K.; Gomez-Gomez, M.M.; Camara, C.; Ramos, L. Silver speciation and characterization of nanoparticles released from plastic food containers by single particle icpms. *Talanta* **2016**, *151*, 83–90. [CrossRef] [PubMed]

168. Hetzer, B.; Burcza, A.; Graf, V.; Walz, E.; Greiner, R. Online-coupling of AF^4 and single particle-ICP-MS as an analytical approach for the selective detection of nanosilver release from model food packaging films into food simulants. *Food Control* **2017**, *80*, 113–124. [CrossRef]

169. McClements, D.J.; Xiao, H.; Demokritou, P. Physicochemical and colloidal aspects of food matrix effects on gastrointestinal fate of ingested inorganic nanoparticles. *Adv. Colloid Interface Sci.* **2017**, *246*, 165–180. [CrossRef] [PubMed]

170. Huang, H.; Tang, K.C.; Luo, Z.S.; Zhang, H.X.; Qin, Y. Migration of Ti and Zn from nanoparticle modified ldpe films into food simulants. *Food Sci. Technol. Res.* **2017**, *23*, 827–834. [CrossRef]

171. Guidance, D. Guidance for industry considering whether an FDA-regulated product involves the application of nanotechnology. *Biotechnol. Law Rep.* **2011**, *30*, 613–616. [CrossRef]

172. Baiguini, A.; Colletta, S.; Rebella, V. Materials and articles intended to come into contact with food: Evaluation of the rapid alert system for food and feed (RASFF) 2008-2010. *Igiene e Sanita Pubblica* **2011**, *67*, 293–305. [PubMed]

© 2018 by the authors. Licensee MDPI, Basel, Switzerland. This article is an open access article distributed under the terms and conditions of the Creative Commons Attribution (CC BY) license (http://creativecommons.org/licenses/by/4.0/).

![nanomaterials logo] *nanomaterials*

MDPI

Article

Active Food Packaging Coatings Based on Hybrid Electrospun Gliadin Nanofibers Containing Ferulic Acid/Hydroxypropyl-Beta-Cyclodextrin Inclusion Complexes

Niloufar Sharif [1], Mohammad-Taghi Golmakani [1], Mehrdad Niakousari [1], Seyed Mohammad Hashem Hosseini [1], Behrouz Ghorani [2] and Amparo Lopez-Rubio [3,*]

[1] Department of Food Science and Technology, School of Agriculture, Shiraz University, km 12 Shiraz-Esfahan Highway, 71441-65186 Shiraz, Iran; sharif1986@shirazu.ac.ir (N.S.); golmakani@shirazu.ac.ir (M.-T.G.); niakosar@shirazu.ac.ir (M.N.); hhosseini@shirazu.ac.ir (S.M.H.H.)

[2] Department of Food Nanotechnology, Research Institute of Food Science and Technology (RIFST), km 12 Mashhad-Quchan Highway, 91895/157/356 Mashhad, Iran; b.ghorani@rifst.ac.ir

[3] Food Quality and Preservation Department, IATA-CSIC, 46980 Paterna, Valencia, Spain

* Correspondence: amparo.lopez@iata.csic.es; Tel.: +34-963-900-022; Fax: +34-963-636-301

Received: 18 October 2018; Accepted: 5 November 2018; Published: 7 November 2018

Abstract: In this work, hybrid gliadin electrospun fibers containing inclusion complexes of ferulic acid (FA) with hydroxypropyl-beta-cyclodextrins (FA/HP-β-CD-IC) were prepared as a strategy to increase the stability and solubility of the antioxidant FA. Inclusion complex formation between FA and HP-β-CD was confirmed by Fourier transform infrared spectroscopy (FTIR), differential scanning calorimeter (DSC), and X-ray diffraction (XRD). After adjusting the electrospinning conditions, beaded-free fibers of gliadin incorporating FA/HP-β-CD-IC with average fiber diameters ranging from 269.91 ± 73.53 to 271.68 ± 72.76 nm were obtained. Control gliadin fibers containing free FA were also produced for comparison purposes. The incorporation of FA within the cyclodextrin molecules resulted in increased thermal stability of the antioxidant compound. Moreover, formation of the inclusion complexes also enhanced the FA photostability, as after exposing the electrospun fibers to UV light during 60 min, photodegradation of the compound was reduced in more than 30%. Moreover, a slower degradation rate was also observed when compared to the fibers containing the free FA. Results from the release into two food simulants (ethanol 10% and acetic acid 3%) and PBS also demonstrated that the formation of the inclusion complexes successfully resulted in improved solubility, as reflected from the faster and greater release of the compounds in the three assayed media. Moreover, in both types of hybrid fibers, the antioxidant capacity of FA was kept, thus confirming the suitability of electrospinning for the encapsulation of sensitive compounds, giving raise to nanostructures with potential as active packaging structures or delivery systems of use in pharmaceutical or biomedical applications.

Keywords: gliadin; ferulic acid; hydroxypropyl-beta-cyclodextrin; electrospinning

---

1. Introduction

Ferulic acid (FA, 4-hydroxy-3-methoxy cinnamic acid), a phenolic compound classified in the group of the hydroxycinnamic acids, is present in commelinid plants such as rice, wheat, oats, and some vegetables, fruits, and nuts [1]. FA exhibits a wide range of biological and biomedical effects including antioxidant [2], anti-inflammatory [3], anti-diabetic [4], hepatoprotective [5] and anti-carcinogenic [6], among others. In addition, FA has been approved as a food additive, hindering the peroxidation of lipids due to its scavenging capability of superoxide anion radicals [7,8]. Despite its biological activity, direct incorporation of FA to foods is limited due to its low water solubility [9] and poor stability under physical and thermal stresses, highlighting its susceptibility to light and oxygen exposure [10]. Interestingly, these limitations might be overcome by the formation of inclusion complexes with cyclodextrins (CDs).

CDs are cyclic, natural, and nontoxic oligosaccharides produced by the enzymatic degradation of starch [11]. Based on the number of α-1,4-linked glucopyranose units in their cyclic structure, different CDs exist, being the most common ones α-CD, β-CD, and γ-CD having 6, 7, and 8 glucopyranose units, respectively [12]. CDs have toroid-shaped molecular structures, a hydrophobic internal cavity and a hydrophilic external surface that make them capable of forming noncovalent host–guest inclusion complexes with a variety of molecules such as phenolic compounds [13]. Therefore, this unique capability offers outstanding improvements in the properties of the guest molecules including protection from degradation and oxidation, enhancing solubility, chemical stability and controlling the release rate [14,15].

Recently, electrospinning, a versatile and cost-effective technique has gained a great interest for fiber fabrication in the range of micron, submicron, and nanoscales [16]. The electrospun fibers have key advantages including high surface-to-volume ratios, small pore sizes and high porosity [17,18]. From an application standpoint, biopolymer-based electrospun fibers can be used for bioactive encapsulation with potential use as delivery systems, controlled release agents or active packaging structures, amongst others [19]. The combination of biopolymers as fiber matrices with organic and inorganic particles (such as clay nanoparticles) for the release of functional molecules has been shown to be an excellent strategy to produce bioactive packaging structures [20–23]. Amongst the natural biopolymers to be used as electrospun matrices, proteins have been widely studied due to their renewable, biodegradable, and biocompatible character [24]. Particularly prolamins, which are plant storage proteins normally obtained as by-products of the starch or beta-glucan production processes, are interesting raw materials for fabricating cost-effective electrospun structures [25], as they are considered more "environmentally economical" when compared with proteins from animal sources [26]. Gliadins, storage prolamins present in wheat kernels, consists of a central domain containing highly repetitive amino acid sequences including proline and glutamine residues and hydrophobic terminal domains which surround the central part. Therefore, gliadin has amphiphilic properties [27]. Gliadins are poorly soluble in aqueous solutions except at extreme pH conditions [28]. It has been found that gliadin has bioadhesive properties, being able to interact with the intestinal membrane through electrostatic interactions and hydrogen bonding [29]. Therefore, gliadins are ideal candidates to fabricate electrospun fiber mats for various applications.

Given the potential of CDs to increase the stability and solubility of hydrophobic bioactives and the excellent properties of gliadin as encapsulation matrix, the aim of the present work was to combine both structures to generate hybrid systems to be used as active coatings for the protection and enhanced solubility of FA within food products. To this end, solid FA/hydroxypropyl-Beta-cyclodextrin inclusion complexes (FA/HP-β-CD-ICs) were first formed, which were subsequently incorporated within gliadin fibers through electrospinning. The inclusion complexes were characterized by X-ray diffraction (XRD), differential scanning calorimeter (DSC), thermogravimetric analysis (TGA) and Fourier transform infrared (FTIR) spectroscopy. Gliadin fibers containing free FA were produced as control samples. The morphological characterization of the hybrid fibers was carried out by scanning electron microscope (SEM). In addition, the presence and distribution of FA in HP-β-CD-ICs and fibers were investigated by fluoresce microscopy. The thermal stability, photostability, release behavior and antioxidant capacity of the developed fibers were also evaluated.

2. Materials and Methods

2.1. Materials

Wheat gluten was purchased from a local shop (Shiraz, Iran). Ferulic acid (Carbosynth, Newbury, England), hydroxypropyl-β-cyclodextrin (Carbosynth, Newbury, England), 2,2-Diphenyl-1-picrylhydrazyl

(DPPH, Sigma-Aldrich, St. Louis, Missouri, United States), phosphate buffer solution (PBS, pH = 7.3 ± 0.1, DNAbiothec, Tehran, Iran), acetic acid (Scharlab, Barcelona, Spain) and ethanol (Panreac, Barcelona, Spain) were purchased and used as received without any further purification. All other chemicals used were of analytical grade unless otherwise specified.

2.2. Preparation of Solid FA/HP-β-CD-IC

The inclusion complexes between FA and HP-β-CD were prepared using the freeze-drying method described by Kfoury, Auezova [30]. FA and HP-β-CD were mixed in aqueous solution in a 1:1 M ratio at a concentration of 10 mM, mixing for 24 h at room temperature. Then, the solution was filtered, frozen and lyophilized by a laboratory freeze dryer (ALPHA 2-4 LD plus, Martin Christ, Osterode am Harz, Germany) at 85 °C and Pa for 48 h. The inclusion ratio (IR%) was calculated using Equation (1):

$$IR\% = (\text{experimental FA content in the solid IC/theoretical FA content}) \times 100 \qquad (1)$$

2.3. Gliadin Extraction

The gliadin fraction of gluten was extracted using the method described by Hong, Trujillo [31] with slight modifications. Briefly, samples of dried gluten powder (20 g) were gently stirred in an ethanol/water mixture (70/30 v/v; gluten/solvent ratio of 1/12) for 4 h at 20 °C. The suspension was centrifuged to collect the gliadin fraction at 10,000 g for 10 min. Finally, the ethanol was evaporated at ambient conditions. The extraction yield of gliadin from wheat gluten powder was 37.5%. The protein content as determined using the Kjeldahl method was 89.8% on a dry matter basis.

2.4. Preparation and Characterization of Gliadin Solutions for Electrospinning

Initially, 25% (w/v) gliadin solutions were prepared by stirring the protein powder in acetic acid at ambient conditions until complete dissolution. Subsequently, FA or FA/HP-β-CD-IC were added into the gliadin solutions (at concentrations of 5, 10 and 20% w/w, with respect to the biopolymer).

The solution properties that affect the electrospinning process, specifically the apparent viscosity, surface tension, and electrical conductivity were evaluated. The surface tension of the solutions was measured using the Wilhemy plate method in an EasyDyne K20 tensiometer (Krüss GmbH, Hamburg, Germany) after calibration of the equipment with deionized water. The electrical conductivity of the solutions was measured using a conductivity meter XS Con6 (Labbox, Barcelona, Spain). The apparent viscosity of the gliadin solutions was determined by a rotational viscometer VISCO BASIC PLUS L from Fungilab S.A. (Sant Feliu de Llobregat, Spain) at 10 rpm using the TL5 spindle. All measurements were made at 25 °C. All Experiments were performed, at least, in triplicate.

2.5. Hybrid Gliadin-Based Fiber Formation Through Electrospinning

Gliadin fibers incorporating free FA (G-FA) and FA/HP-β-CD-IC (G-FA/HP-β-CD-IC) were fabricated via electrospinning. The electrospinning process was conducted using an electrohydrodynamic apparatus equipped with a variable high voltage 0–35 kV power supplier (spinner-3X-Advance, ANSTCO, Tehran, Iran). Solutions were loaded into 10 mL disposable plastic syringes and the electrospinning process was conducted at the voltage of 18 kV and a flow rate of 1 mL/h. Tip to collector distance was kept constant at 100 mm. The obtained fibers were collected on aluminum foil attached to the surface of the collector and kept overnight under the hood to evaporate any solvent residues. All experiments were performed at ambient conditions.

2.6. Ultraviolet-Visible Spectroscopy

The Ultraviolet-Visible (UV-Vis) spectra of FA, HP-β-CD and their corresponding inclusion complexes were recorded on a UV-Vis spectrophotometer (UV-1280, Shimadzu Corporation,

Kyoto, Japan). Each sample (0.3 mM) was dissolved -in methanol and the spectra were obtained in the range from 220 to 400 nm.

2.7. Optical and Scanning Electron Microscopy (SEM)

The morphology of electrospun gliadin structures containing FA and FA/HP-β-CD-IC was examined by a TESCAN-Vega 3 scanning electron microscope (SEM) (TESCAN, Brno, Czech Republic). SEM was conducted at an accelerating voltage of 20 kV and at working distances of 9-16 mm after sputter coating the electrospun webs with gold under vacuum (Q 150R-ES; Quorum Technologies, Laughton, UK). Image analysis software (Digimizer, MedCalc Software, Ostend, Belgium) was used to determine fiber diameters from the SEM micrographs in their original magnification. Average fiber diameters (AFD) and fiber size distributions were obtained from a minimum of 100 measurements. The presence and distribution of FA in HP-β-CD inclusion complex and fibers were investigated using a digital microscopy system (Nikon Eclipse 90i, Barcelona, Spain) fitted with a 12 V, 100 W halogen lamp and equipped with a digital imaging head which integrates an epifluorescence illuminator. A digital camera head (Nikon DS-5Mc, Tokyo, Japan) with a 5-megapixel CCD cooled with a Peltier mechanism was attached to the microscope.

2.8. Encapsulation Efficiency

Encapsulation efficiency (EE%) was calculated by measuring the non-entrapped FA according to Yang, Feng [32] with some modifications. Briefly, Fibers (10–20 mg) was submerged in absolute ethanol (8 mL) for 30 s. Then the mixture was centrifuged at 2500 rpm for 10 min and the absorbance of FA was then determined by UV-Vis spectrophotometer (UV-1280, Shimadzu Corporation, Kyoto, Japan) based on the calibration curve ($R^2 = 0.999$) obtained for FA in absolute ethanol at a wavelength of 310 nm. The EE% values were calculated using Equation (2):

$$EE\% = ((\text{total theoretical mass of FA-free mass of FA in the mixture})/\text{total theoretical mass of FA}) \times 100 \quad (2)$$

2.9. Fourier Transform Infrared (FTIR) Spectroscopy

The infrared spectra of pure FA, pure HP-β-CD, FA/HP-β-CD-IC, gliadin fiber, G-FA fiber, and G-FA/HP-β-CD-IC fibers were investigated using a Fourier transform infrared (FTIR) spectrometer (model FTIR-8400S, Shimadzu Corp., Kyoto, Japan). The scans were done in the mid-infrared region in the range of 4000–400 cm^{-1} wavenumber at a spectral resolution of 2 cm^{-1}.

2.10. Differential Scanning Calorimetry (DSC) and Thermogravimetric Analyze (TGA)

Thermal properties of pure FA, pure HP-β-CD, FA/HP-β-CD-IC, gliadin fiber, G-FA fiber, and G-FA/HP-β-CD-IC fibers were investigated by differential scanning calorimetry (DSC) (PerkinElmer, Akron, OH, USA) and thermogravimetric analysis (TGA) (TA Instruments, New Castle, DE, USA). The DSC analyses were conducted within a temperature range from 35 °C to 250 °C at a heating rate of 10 °C/min under N_2 gas flow at a flow rate of 50 mL/min. The TGA measurements were carried out from 25 to 700 °C at 10 °C/min heating rate under N_2 flow of 20 mL/min as a purge gas for both the balance and the sample.

2.11. X-ray Diffraction (XRD)

The crystalline structure of FA, HP-β-CD, and FA/HP-β-CD-IC was investigated using an X-ray diffractometer (model D8-ADVANCE, Bruker, Germany) with Cu Kα radiation. The samples were examined over the angular range of 2θ 5°–80°.

2.12. Antioxidant Activity

The antioxidant activity of free FA, G-FA, and G-FA/HP-β-CD-IC electrospun fibers was determined using the 2,2-diphenyl-1-picrylhydrazyl (DPPH) radical scavenging assay at various FA concentrations [33]. The fiber mats having the equivalent amount of FA were immersed in 2 mL of water and stirred. 2mL of 10^{-4} M DPPH solution in methanol were added to the previous solutions. The absorbance of the solutions was measured by UV-Vis spectrophotometer (UV-1280, Shimadzu Corporation, Kyoto, Japan) after 60 min. The antioxidant activity of samples was determined as:

$$\text{Antioxidant activity (\%)} = [(A_{control} - A_{sample})/A_{control}] \times 100 \tag{3}$$

where $A_{control}$ and A_{sample} are the absorbances of DPPH solution without sample and DPPH solution with the sample (Free FA or hybrid gliadin fibers), respectively.

2.13. Photostability

The photostability of FA incorporated within the fibers either in free form or as an inclusion complex was evaluated. Briefly, solid fibers were cut into square-shaped samples and positioned 7 cm away from a UV light source (75 W at 253.7 nm, Model NIQ 80/36 U, Heraeus, Boadilla del Monte, Madrid, Spain) in a chamber at ambient conditions. At different time intervals (0, 15, 30 and 60 min), the remaining amounts of FA were measured by UV-Vis spectroscopy at 310 nm after immersing samples into 70% ethanol solution. For comparison purposes, the free FA solution was also investigated. Each sample was analyzed at least in triplicate and the results were expressed as an average ± standard deviation.

2.14. In Vitro Release Assays

The in vitro release studies were carried out for selected fibers in two different media: 10% ethanol as food simulant for aqueous food products and 3% acetic acid as acidic food products [34]. In addition, we also investigated the release in PBS aqueous buffer as a biological fluid simulant. A method adapted from Atay, Fabra [35] was used for that means. Briefly, 10 mg of fibers were incorporated into 10 mL of media at ambient conditions. At specified time intervals, the samples were centrifuged at 2000 rpm for 2 min (Eppendorf centrifuge 5804r, Hamburg, Germany). Then, 1 mL aliquot of supernatant was withdrawn for analysis, replacing with fresh release medium and re-suspending. Finally, the concentration of FA in release media was calculated by measuring the absorbance of the supernatant at a wavelength of 310 nm using a UV-Vis spectrophotometer (UV-1280, Shimadzu Corporation, Kyoto, Japan). Three independent replicates of each fiber were carried out and the results were reported as average ± standard deviation.

2.15. Statistical Analyses

The obtained data was expressed as the mean ± standard deviation of triplicate determinations. Statistical significance among treatments were evaluated with analysis of variance (one-way ANOVA with Tukey's post hoc test), using SPSS 25 (SPSS Inc., Chicago, IL, USA) statistical software. Tukey's multiple range tests were applied to determine the significance of differences between mean values ($p < 0.05$).

3. Results and Discussion

3.1. Preparation of FA/HP-β-CD-IC

In recent years, CDs have gained increased attention to develop different guest-host complexes in order to improve solubility, stability, and bioavailability of a wide range of compounds [9]. A number of techniques have been developed to prepare CD-IC including co-precipitation, kneading, spray-drying and freeze-drying, among others [36]. The freeze-drying method is attracting more and more attention due to its advantages such as protection against chemical decomposition, minimal effect on guest compound activity due to processing at low temperatures as well as low moisture content amount at final physical IC [9]. Hence, we prepared FA/HP-β-CD-IC by the freeze-drying method. The content of FA in FA/HP-β-CD-IC was 11.06 ± 0.22% and the inclusion ratio of FA was 89.49 ± 0.66%. The inclusion ratio was higher than that reported by previous works [9,37], which could be ascribed to the different method of IC preparation.

3.2. Characterization of the Inclusion Complexes

XRD analysis has been said to be a useful technique to confirm the formation of inclusion complexes. Apparently, once the guest molecules are within the CD cavities, the crystalline peaks of guest molecules cannot be detectable [38]. Several intense and sharp diffraction peaks at 2 theta values around 9°, 10°, 16°, 17°, 27° and 29° were observed for FA (Figure S1 from the Supplementary Material), indicating its crystalline nature [9]. In contrast, HP-β-CD displayed an amorphous halo, confirming its amorphous character. No characteristic peaks were observed in the case of FA/HP-β-CD-IC, which could indicate the successful formation of the guest-host inclusion complexes as suggested by previous studies.

Generally, when guest molecules form inclusion complexes with cyclodextrin molecules, they might exhibit different characteristics than the pure compounds [39]. For instance, thermal transitions including melting, boiling or sublimation temperatures have been observed to shift to a different temperature or disappear [20,40]. Therefore, the DSC thermograms of FA, HP-β-CD and FA/HP-β-CD-IC were obtained. As shown in Figure 1a, the DSC thermogram of pure FA displayed a sharp endothermic peak at around 176 °C, which corresponds to its melting point, followed by a broad peak at higher temperatures attributed to decomposition. In the case of HP-β-CD, an amorphous molecule, there was a broad endothermic peak at about 84 °C, corresponding to the release of water [39]. After the formation of the inclusion complex, a different thermogram was observed. The absence of the melting point of FA and shifting of the characteristic peak of HP-β-CD (from 84 °C to 67 °C), both seemed to indicate that FA was successfully included into the cavity of HP-β-CD during the formation of the inclusion complex (Figure 1b). It has been extensively reported that when inclusion complexes with CDs are formed, the guest molecules lose their characteristic peaks in the DSC thermograms [9,37].

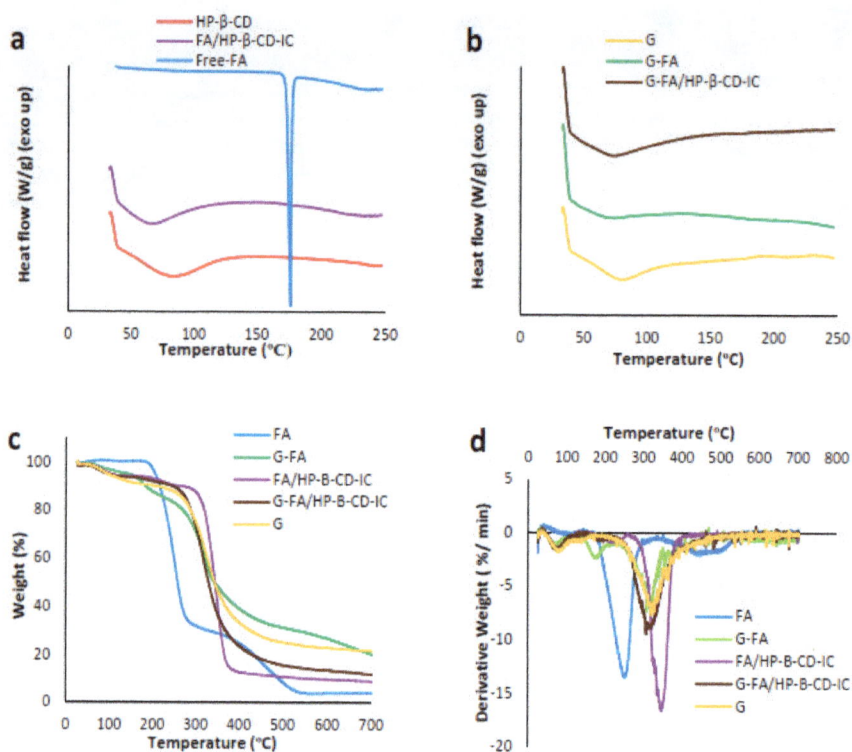

Figure 1. (a,b) differential scanning calorimeter (DSC) thermograms of pure ferulic acid (FA), HP-β-CD and FA/HP-β-CD-IC and Gliadin, G-FA 20%, and G-FA/HP-β-CD-IC 20% electrospun fibers, respectively. (c) Thermogravimetric analysis (TGA) and (d) Derivative thermogravimetric (DTG) curves of pure FA, and FA/HP-β-CD-IC, Gliadin, G-FA 20%, and G-FA/HP-β-CD-IC 20% electrospun fibers.

UV-Vis spectroscopy was also used for the characterization of the inclusion complexes. In this study, the UV-Vis spectra were recorded for FA, HP-β-CD and their inclusion complexes (Figure S2 in the Supplementary Material). HP-β-CD showed a very low UV absorbance without any characteristic absorption peaks. FA exhibited three characteristic peaks at 217, 287 and 310 nm, corresponding to π-π* transition of the phenyl ring, π-π* transition of the phenolic group and π-π* transition of the double bond, respectively [9]. On the other hand, in the spectrum of FA/HP-β-CD-IC, these characteristic peaks had slightly shifted (almost 2 nm to higher absorbance) suggesting the presence of non-covalent interactions between FA and HP-β-CD [41].

According to Ram, Seitz [42], FA is known to have a blue emission fluorescence wavelength at around 425 nm. Therefore, this characteristic was used to visually confirm that FA had been effectively incorporated into the HP-β-CD. As observed in Figure 2, HP-β-CD did not have any fluorescence emission while the inclusion complex with FA endowed its fluorescence properties due to the presence of FA, thus further confirming the successful IC formation.

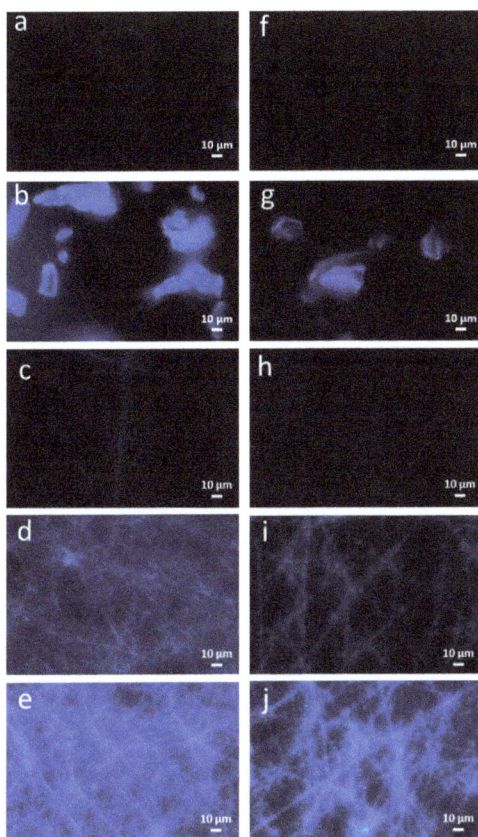

Figure 2. Fluorescence microscopy images of electrospun structures: (**a**) pure gliadin; (**b**) pure FA; (**c**) G-FA 5%; (**d**) G-FA 10%; (**e**) G-FA 20%; (**f**) pure HP-β-CD; (**g**) FA/HP-β-CD-IC; (**h**) G-FA/HP-β-CD-IC 5%; (**i**) G-FA/HP-β-CD-IC 10%; (**j**) G-FA/HP-β-CD-IC 20%.

FTIR spectroscopy was used as an additional tool to confirm the formation of a host-guest inclusion complexes. The FTIR spectra of pure FA, pure HP-β-CD, and FA/HP-β-CD-IC are depicted in Figure 3a. The FTIR spectrum of FA has characteristic peaks in the region of 3435 cm^{-1} (O–H stretching vibration), 1689, 1663, and 1618 cm^{-1} (C=O stretching vibration), 1590, 1517, and 1431 cm^{-1} (aromatic skeleton vibration). Absorption at 1466 cm^{-1} arises from single bond C–H deformations and aromatic ring vibrations while absorption at 1276 cm^{-1} is attributed the C–O–C asymmetric stretching vibration. The peak in the region of 1176 cm^{-1} is characteristic of the carbonyl group. Moreover, the bands at 852 and 804 cm^{-1} are related to the two adjacent hydrogen atoms on the phenyl ring in the FA structure [9,32]. The FTIR spectrum of HP-β-CD exhibited prominent absorption bands located at 3408 cm^{-1} (O-H stretching vibrations), 2924 cm^{-1} (C–H stretching vibrations), and 1127 and 1036 (C–H and C–O stretching vibrations) [9,43]. Several spectral changes were observed upon formation of the FA/HP-β-CD-IC, especially in the region from 1000 to 1900 cm^{-1}. Comparing the HP-β-CD spectrum with that from the inclusion complex, new bands, probably arising from the incorporation of FA in the structure were observed at 1676 cm^{-1} and 1100 cm^{-1} (see arrows in Figure 3a). In addition, the characteristic peaks from the carbon stretching vibrations from the HP-β-CD shifted towards greater wavenumbers, probably due to conformational changes taking place as a consequence of incorporating the FA molecules within the CD structure.

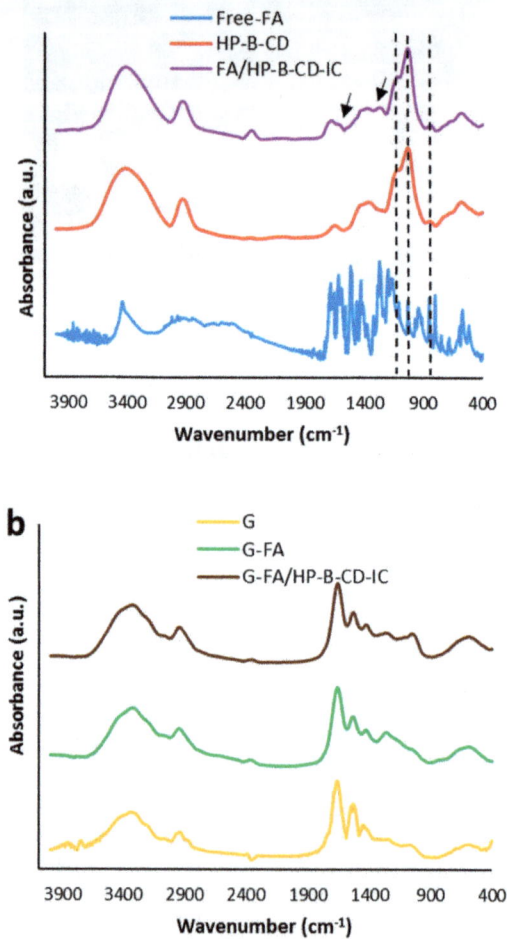

Figure 3. Fourier transform infrared (FTIR) spectra of (**a**) pure FA, HP-β-CD and FA/HP-β-CD-IC; (**b**) Gliadin, G-FA 20%, and G-FA/HP-β-CD-IC 20% electrospun fibers.

3.3. Solution Properties, Fibers Morphology, and Distribution of FA Within the Fibers

In order to better understand how solution properties affected the morphology of gliadin in the presence of FA, three different concentrations of FA and FA/HP-β-CD-IC (5, 10 and 20% w/v with respect to biopolymer) were investigated. Table 1 compiles these solution properties: viscosity, electrical conductivity, and surface tension. The viscosity of G-FA and G-FA/HP-β-CD-IC solutions at all studied concentrations was higher than that of pure gliadin solutions possibly due to the interactions between the gliadin biopolymer chains and FA. On the other hand, G-FA and G-FA/HP-β-CD-IC solutions at all FA studied concentrations exhibited lower conductivity than that of the pure gliadin solution. Moreover, no significant difference was observed for the surface tension of the different solutions except for the solutions with the greatest FA concentration.

Table 1. Solution concentrations and properties used in electrospinning process.

Solution	%Gliadin[1] (w/w)	%HP-β-CD[2] (w/w)	%Ferulic acid[2] (w/w)	%Ferulic acid/HP-β-CD IC[2] (w/w)	Viscosity[3] (mPa.s 10 rpm, 25 °C)	Surface Tension[3] (mN.m^{-1})	Electrical Conductivity[3] (μs.cm^{-1})
Gliadin	25	-	-	-	197.9 ± 1.61 ^{e}	28.80 ± 0.37 ^{b}	148.25 ± 0.78 ^{a}
Gliadin/Ferulic acid	25	-	5	-	219.1 ± 2.71 ^{d}	29.05 ± 0.08 ^{ab}	89.70 ± 1.73 ^{bc}
	25	-	10	-	238.3 ± 2.44 ^{bc}	29.30 ^{ab}	90.17 ± 0.93 ^{bc}
	25	-	20	-	247.2 ± 3.92 ^{b}	29.80 ± 0.14 ^{a}	81.1 ± 0.71 ^{d}
Gliadin/Ferulic acid/HP-β-CD IC	25	-	-	5	219.1 ± 7.56 ^{d}	29.35 ± 0.35 ^{ab}	93 ± 0.85 ^{b}
	25	-	-	10	244.4 ± 7.37 ^{b}	29.55 ± 0.08 ^{ab}	87.50 ± 3.39 ^{c}
	25	-	-	20	277.1 ± 9.43 ^{a}	29.45 ± 0.35 ^{ab}	89.03 ± 1.58 ^{bc}

1—With respect to the solvent (acetic acid). 2—With respect to the polymer (gliadin). 3—Data are displayed in means ± standard deviation of three replications ($p < 0.05$); means in each column bearing different superscripts are significantly different ($p < 0.05$).

The morphology of gliadin, G-FA and G-FA/HP-β-CD-IC electrospun fibers was investigated using scanning electron microscopy (SEM). Representative SEM images of the fiber mats and average fiber diameters (AFD) along with fiber distributions are given in Figure 4. In our previous work, the conditions for electrospinning pure gliadin fibers were optimized and uniform and beaded-free fibers having AFD 256.49 ± 78.49 nm were obtained [44]. Incorporation of free and complexed FA led to the formation of slightly thicker fibers, explained by the greater viscosity of the electrospinning solutions. Moreover, the AFD of gliadin-FA fibers was slightly higher than gliadin-FA/HP-β-CD-IC fibers. These slight variations in average fiber diameter (AFD) for fibers were most likely due to differences in solution properties such as viscosity and electrical conductivity after incorporating the bioactive compound [45,46]. As explained above, G-FA and G-FA/HP-β-CD-IC solutions at all studied concentrations had higher viscosity and lower conductivity than pure gliadin solutions. Generally, solutions with higher viscosity and lower conductivity result in thicker fibers as less stretching of the jet occurs during the electrospinning process [45,46]. In addition, the thickest fibers (279.42 ± 80.85 nm) were obtained from G-FA/HP-β-CD-IC solutions with the greatest FA content (20% *w/w* with respect to the polymer), which had lower conductivity and higher surface tension, thus contributing to less stretching of the electrified jet as a result of less repulsion of charges on the surface during electrospinning process [14,47]. But, in general, uniform and beaded free fibers with almost similar AFD were successfully fabricated from G-FA and G-FA/HP-β-CD-IC solutions at different FA concentrations.

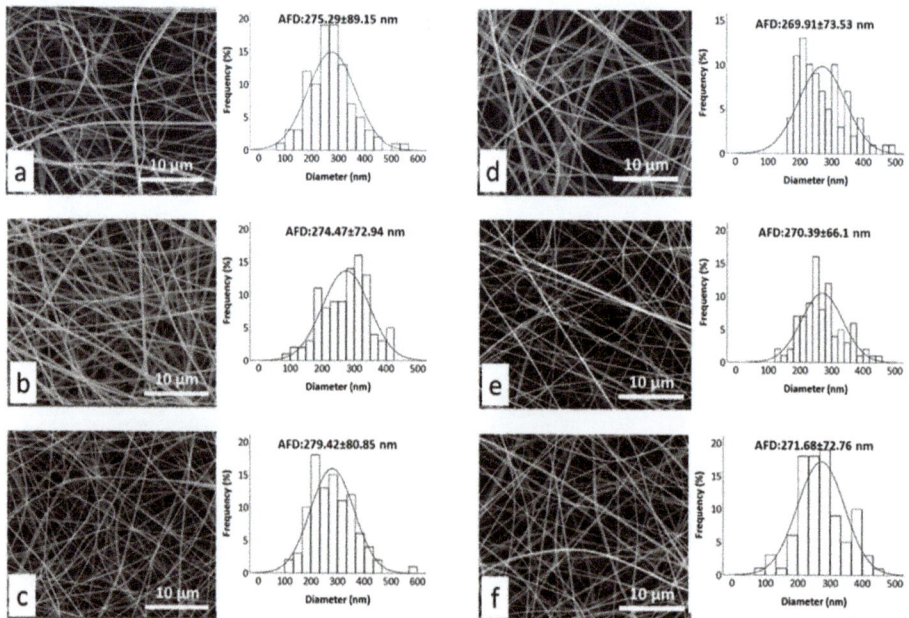

Figure 4. Representative scanning electron microscope (SEM) images and average fiber diameter (AFD) of electrospun structures: (**a**) G-FA 5%; (**b**) G-FA 10%; (**c**) G-FA 20%; (**d**) G-FA/HP-β-CD-IC 5%; (**e**) G-FA/HP-β-CD-IC 10%; (**f**) G-FA/HP-β-CD-IC 20%.

Fluorescence microscopy was also used to study the distribution of FA along the fibers. As observed in Figure 2 pure gliadin fibers did not have any intrinsic fluorescence, while the hybrid fibers displayed a rather homogeneous color, suggesting that FA was effectively distributed along the fibers. It was also observed that increasing the amount of bioactive compound led to enhanced blue color intensity, confirming higher loading efficiencies at higher bioactive concentrations. The intensity of the blue color in the gliadin fibers containing the free FA was higher than in the fibers with the inclusion complexes, explained by the greater bioactive concentration in the hybrid structures.

3.4. Encapsulation Efficiency

The encapsulation efficiency (EE%) of FA in G-FA and G-FA/HP-β-CD-IC electrospun fibers at different concentrations were calculated by Equation 2. The data indicated that almost 100% FA was effectively incorporated into the gliadin electrospun fibers. The EE% of 5%, 10% and 20% FA-loaded fibers were $97.05 \pm 0.66\%$, $95.09 \pm 0.98\%$ and $94.03 \pm 4.88\%$, respectively. The EE% of 5%, 10% and 20% FA/HP-β-CD-IC-loaded fibers were $95.30 \pm 1.66\%$, $96.93 \pm 0.70\%$ and $95.65 \pm 1.38\%$, respectively. These values are higher than those reported for the encapsulation of FA with other electrospun biopolymers including amaranth protein isolate and pullulan [10] or using other encapsulation techniques [4,48]. Therefore, the obtained results suggest that gliadin electrospun mats can be used as an efficient encapsulant for bioactive food compounds such as FA. Given the excellent encapsulation capability of the fibers, irrespective of the added FA, the materials with the greatest amount of free FA and FA/HP-β-CD-IC (i.e., 20% with respect to the polymer) were selected for further characterization.

3.5. Infrared Analysis of the Electrospun Fibers

Figure 3b shows the infrared spectra of pure gliadin fibers and the hybrid electrospun fibers containing either free FA or the inclusion complexes. The spectrum of pure gliadin fibers is characterized by the bands at 1660 and 1540 cm^{-1} attributed to the C=O and C–N stretching vibration (Amide I) and N–H bending vibration and C–N and C–C stretching vibration (Amide II), respectively. Upon incorporation of free FA or the inclusion complexes, the amide I band from the fibers shifted to 1658 and 1655 cm^{-1}, respectively. Similarly, a shift in the amide II band from 1540 to 1531 cm^{-1} was also observed for both hybrid fibers containing the bioactive compound. These results revealed that the incorporated FA and inclusion complexes were interacting with the amino groups from the prolamin [49]. According to Torres-Giner, Gimenez [50], the frequencies of amide I, and II reflects the size of the α-helix structure in the biopolymer; a shift toward lower wavenumbers suggests greater structural stability, which is directly related to an increase hydrogen bonding interactions taking place through the N–H groups from the protein. The band shifts observed in the FTIR spectra of G-FA and G-FA/HP-β-CD-IC electrospun fibers indicated that there were interactions among gliadin and FA or FA/HP-β-CD-IC, altering the secondary structure of the prolamin, especially when incorporating FA into gliadin electrospun fibers. In addition, there were hydrophobic interactions between the hydrophobic residue of the prolamin and phenyl ring of FA [51], which was absent in the case of incorporating FA/HP-β-CD-IC into gliadin electrospun fibers as the aromatic ring of FA in the latter case was involved in the formation of the inclusion complex.

3.6. Thermal Properties of the Electrospun Fibers

DSC was carried out to investigate the thermal properties of electrospun pure gliadin, G-FA, and G-FA/HP-β-CD-IC fibers. The DSC thermogram of the pure gliadin fiber exhibited a single endothermic at 80 °C, which has been normally termed as dehydration temperature (T_d), corresponding the loss of bound water from the material [52]. The DSC thermograms of FA and FA/HP-β-CD-IC after incorporation into gliadin fibers were similar to gliadin fiber, showing a slight shift to around 74 °C and 73 °C, respectively; i.e., the high surface area of the generated materials facilitated water evaporation. Moreover, in the case of FA, the obtained DSC thermogram did not show the melting peak of pure FA, probably due to the previous dissolution of the bioactive for incorporation within the gliadin structures, which resulted in the crystallinity loss of the compound.

The thermal stability of FA encapsulated in the gliadin fibers in free and inclusion complex form was also investigated. The TGA studies of pure FA and gliadin fibers were also performed for comparison purposes. Table S1 from the Supplementary Material summarizes the main results.

In general, the weight loss in first region (less than 160 °C) is usually due to the water evaporation and volatile compounds while the temperature at which the highest rate of weight loss occurs (i.e., the peak in the derivative thermogram (DTG)) is regarded as degradation temperature (T_d) [53]. Regarding the weight loss related to water evaporation (first thermal transition in the TGA curves), these could be related to the DSC curves, which explain that as very little water was weakly sorbed to the pure ferulic acid compound, the initial stability of the antioxidant molecule was greater than the other materials analyzed (refer to Figure 1). As shown in Figure 1c,d, while the first stage of water evaporation was not observed for the pure antioxidant, degradation of FA occurred in two different stages as previously observed in the literature [54], the first one corresponding to the formation of 4-vinylguaiacol (with a T_d around 250 °C) and the second one mainly related to the formation of unsubstituted, 4-methyl-, and 4-ethylguaiacols at the expense of 4-vinylguaiacol [55]. After the formation of the inclusion complex with HP-β-CD, the main degradation step shifted to higher temperatures (349 °C). The enhanced thermal stability of the guest molecules within inclusion complexes has been reported for other CD-IC systems [56]. The weight loss of neat gliadin fibers took place around 320 °C. Electrospun gliadin fibers incorporating free FA exhibited a two-step degradation, at ~180 and ~310 °C, which might correspond to the T_d of free FA and gliadin, respectively. This decrease in the thermal stability of the antioxidant molecule upon incorporation in protein-based matrices has been previously observed [10]. On the other hand, for G-FA/HP-β-CD-IC fibers, there was only one main degradation step with a maximum around 309 °C, indicating that the formation of inclusion complexes effectively stabilized the bioactive molecules, even though incorporation of the inclusion complexes within the electrospun gliadin fibers seemed to be somehow detrimental in terms of thermal stability (if compared with the excellent thermal stability of the isolated ICs).

3.7. Antioxidant Activity

The antioxidant activity of FA (powder) and after incorporation into gliadin electrospun fibers in the free and IC forms was calculated via DPPH radical scavenging assay. The antioxidant activity of FA, G-FA, and G-FA/HP-β-CD-IC fibers were 92.06 ± 1.06% 91.31 ± 0.56% and 88.79 ± 0.74% to, respectively. The results revealed that incorporation of FA within the fibers (either in free form or as an inclusion complex) did not significantly affect its antioxidant activity. FA preserved its antioxidant activity after incorporation into gliadin fibers in the free form despite the interactions that took place between FA and gliadin after electrospinning in accordance with previous studies in which similar results were obtained for other phenolic acids and prolamins [56]. In addition, the formation of the inclusion complex between FA and HP-β-CD had no effect on the antioxidant activity of FA, which might due to the high solubility of FA/HP-β-CD-IC. Aytac, Ipek [33] also reported no significant differences between the antioxidant activity of quercetin (another phenolic compound), in free form and IC form incorporated into zein fibers. Moreover, the application of high voltage during the electrospinning process had no negative effect on the antioxidant activity of FA since G-FA and G-FA/HP-β-CD-IC fibers had similar antioxidant capacity than the free FA.

3.8. Photostability Analyses

In general, FA might undergo photodegradation upon UV irradiation, causing trans-cis isomerization [57]. It has been reported that CDs might provide protection for their guest compounds against UV irradiation [58,59]. Hence, the photostability of the hybrid fibers was investigated. As shown in Figure 5, upon UV exposure, extensive degradation of free FA occurred, only remaining about 9% of the compound after 60 min irradiation. In contrast, incorporation within the electrospun gliadin fibers, effectively prevented its photodegradation. After 60 min of UV irradiation, the percentage of remaining FA in G-FA fibers was 43% while for G-FA/HP-β-CD-IC fibers it was 76%. In addition, the rate of FA degradation was slower for G-FA/HP-β-CD-IC fibers. It has been reported that the photostability of FA can be improved by the formation of IC with CDs including α-CD [59] and HP-β-CD [9]. In our study, the gliadin electrospun fibers provided a second shield, offering more FA stability blocking the passage of UV light towards the photosensitive compound. Therefore, FA/HP-β-CD-IC incorporated into gliadin fibers improved the photostability of FA, making it less sensitive to UV light.

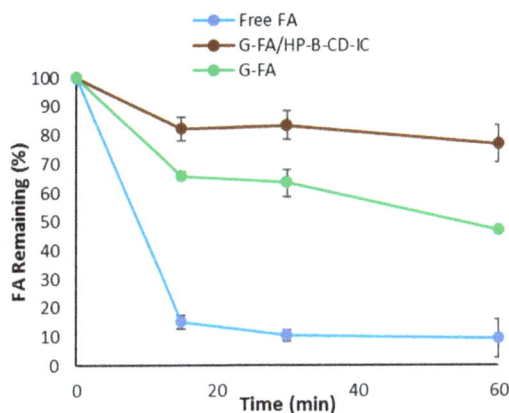

Figure 5. Photodegradation profiles of pure FA, G-FA 20% and G-FA/HP-β-CD-IC 20% electrospun fibers.

3.9. In Vitro Release Assays

The release of FA from gliadin-FA and gliadin-FA/HP-β-CD-IC electrospun fibers in two media, 10% ethanol (as an aqueous food simulant), and 3% acetic acid (as an acidic food simulant) was studied. The obtained release profiles are depicted in Figure 6. The release of FA from G-FA and G-FA/HP-β-CD-IC electrospun fibers showed a similar behavior in acidic medium. The release of FA reached a steady state following an initial burst release that occurred during the first 10 min due to complete dissolution of both fibers in this media, which might be attributed to the greater solubility of gliadin in acetic conditions. In contrast, in the aqueous food simulant, G-FA/HP-β-CD-IC fibers quickly dissolved while swelling of the G-FA fibers occurred. Again, a fast release at the initial stage (75%) was observed for G-FA/HP-β-CD-IC fibers, while a much lower release was observed for the G-FA fibers during the first 10 min (28%), subsequently exhibiting a longer sustained release profile. The difference in the amount and rate of FA release from the fibers was probably due to the different solubility of FA in the different media. It should be highlighted that in both food simulants the release of FA from G-FA/HP-β-CD-IC fibers was greater due to the solubility enhancement in the form of an inclusion complex. It has been reported that HP-β-CD would enhance the solubility of phenolic acids including gallic acid [38]. Moreover, as explained in Section 3.5, there were hydrophobic interactions between the hydrophobic residue of the prolamin and phenyl ring of FA that were absent in the case of incorporating FA/HP-β-CD-IC into gliadin electrospun fibers as the aromatic ring of FA was involved in the formation of the inclusion complex.

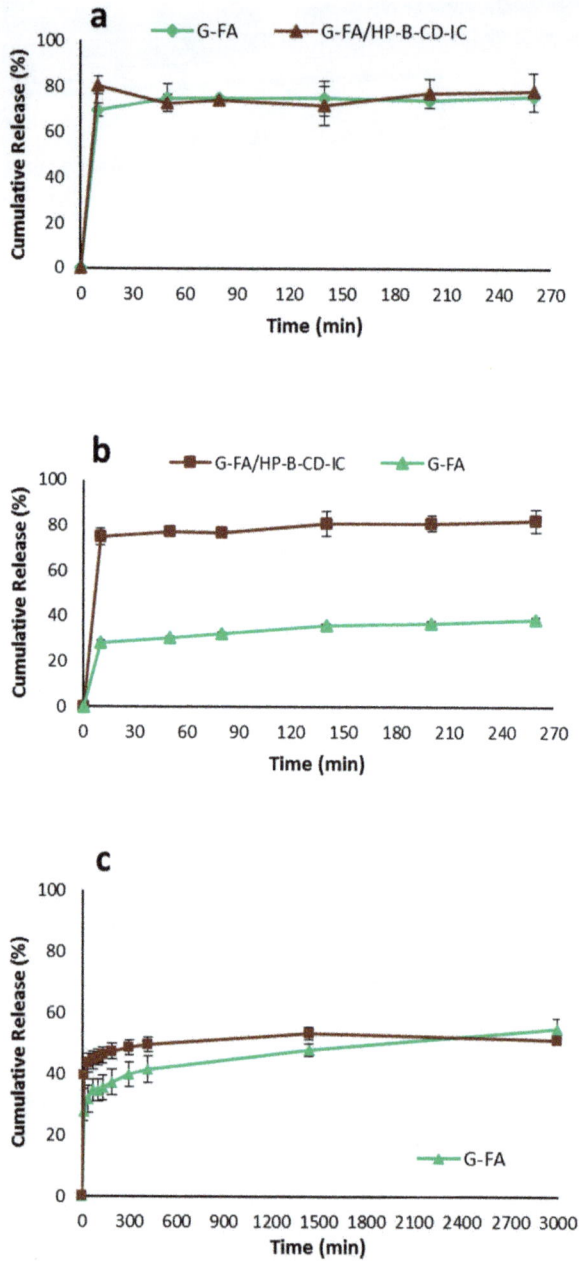

Figure 6. Release behavior of FA from G-FA 20% and G-FA/HP-β-CD-IC 20% electrospun fibers in different media: (**a**) acetic acid 3%; (**b**) ethanol 10%; and (**c**) PBS.

In addition, the release of FA in PBS aqueous buffer as one of the most widely studied blood plasma simulant was also studied. The release FA from G-FA and G-FA/HP-β-CD-IC electrospun fibers can be divided in two stages: an initial fast release and then a slow release. The release of FA from G-FA/HP-β-CD-IC electrospun fibers was higher than G-FA fibers during the initial fast release stage. In other words, G-FA/HP-β-CD-IC electrospun fibers released almost 40% of theoretically loaded FA into PBS medium within 10 min, whereas G-FA electrospun fibers released around 27% of theoretically loaded FA over the same period. Moreover, the release of FA from G-FA/HP-β-CD-IC fibers increased gradually, reaching to a plateau (steady state) after around 7 h. In contrast, a small percentage of the theoretical FA loading was released from G-FA fibers during the same period of time. This again confirms that HP-β-CD effectively improves the solubility of FA. Additionally, HP-β-CD promoted the diffusion of water into fiber mat as a hydrating agent, increasing its porosity [60]. Thus, it could be concluded that the incorporation of FA in the form of FA/HP-β-CD-IC could be used to provide a quick solubility while requiring less amounts of FA due to the high solubility of FA in this case. On the contrary, G-FA fiber mats might serve as matrices for more sustained release applications.

4. Conclusions

In this work, FA was successfully incorporated into gliadin fibers in form of FA/HP-β-CD-IC via the electrospinning technique. First of all, the inclusion complex between FA and HP-β-CD was prepared at a 1:1 molar ratio using a freeze-drying method. The formation of the inclusion complex between FA and HP-β-CD was confirmed by FTIR, DSC, and XRD analyses. Then, FA/HP-β-CD-IC was incorporated into gliadin fibers in order to fabricate hybrid electrospun fibers. Gliadin fibers incorporating FA in free form were also electrospun for comparative purposes. The uniform and beaded-free fibers and FA distribution along the fibers were observed using SEM and fluorescence microscopy, respectively. The obtained electrospun fibers maintained their antioxidant activities in spite of the high voltage applied during the electrospinning process. Moreover, the photostability of FA was significantly improved when incorporating the bioactive in the form of FA/HP-β-CD-IC. The inclusion complexes also favored the solubility of FA in different media. Hence, these fibers could find certain applications in various areas including food, packaging, health, pharmaceutical, among others.

Supplementary Materials: The following are available online at http://www.mdpi.com/2079-4991/8/11/919/s1, Figure S1: X-ray diffraction (XRD) diffraction patterns of pure FA, HP-β-CD and FA/HP-β-CD-IC, Figure S2: UV-visible absorption spectra of pure FA, HP-β-CD and FA/HP-β-CD-IC, Table S1: Temperatures of maximum degradation rate and corresponding weight losses of the different degradation stages and residual matter at 700 °C from the various samples.

Author Contributions: N.S. conducted the experimental part and writing of the first draft. M.-T.G., M.N., S.M.H.H. and B.G. were responsible of the conceptualization and funding acquisition. A.L.-R. was in charge of experimental supervision, data analysis, validation and reviewing and editing the final version of the manuscript.

Acknowledgments: The authors would like to thank the Spanish MINECO project AGL2015-63855-C2-1 for financial support. The authors would also like to thank the Central Support Service for Experimental Research (SCSIE) of the University of Valencia for the electronic microscopy service. In addition, the authors gratefully acknowledge the financial support of the Research Affairs Office at Shiraz University (Grant #93GCU1M1981).

Conflicts of Interest: The authors declare no conflict of interest.

References

1. Doss, H.M.; Samarpita, S.; Ganesan, R.; Rasool, M. Ferulic acid, a dietary polyphenol suppresses osteoclast differentiation and bone erosion via the inhibition of RANKL dependent NF-κB signalling pathway. *Life Sci.* **2018**, *207*, 284–295. [CrossRef] [PubMed]
2. Vashisth, P.; Kumar, N.; Sharma, M.; Pruthi, V. Biomedical applications of ferulic acid encapsulated electrospun nanofibers. *Biotechnol. Rep.* **2015**, *8*, 36–44. [CrossRef] [PubMed]
3. Akihisa, T.; Yasukawa, K.; Yamaura, M.; Ukiya, M.; Kimura, Y.; Shimizu, N.; Arai, K. Triterpene alcohol and sterol ferulates from rice bran and their anti-inflammatory effects. *J. Agric. Food. Chem.* **2000**, *48*, 2313–2319. [CrossRef] [PubMed]

4.   Panwar, R.; Raghuwanshi, N.; Srivastava, A.K.; Sharma, A.K.; Pruthi, V. In-vivo sustained release of nanoencapsulated ferulic acid and its impact in induced diabetes. *Mater. Sci. Eng. C* **2018**, *92*, 381–392. [CrossRef] [PubMed]

5.   Rukkumani, R.; Aruna, K.; Suresh Varma, P.; Padmanabhan Menon, V. Hepatoprotective role of ferulic acid: A dose-dependent study. *J. Med. Food* **2004**, *7*, 456–461. [CrossRef] [PubMed]

6.   Vashisth, P.; Sharma, M.; Nikhil, K.; Singh, H.; Panwar, R.; Pruthi, P.A.; Pruthi, V. Antiproliferative activity of ferulic acid-encapsulated electrospun PLGA/PEO nanofibers against MCF-7 human breast carcinoma cells. *3 Biotech* **2015**, *5*, 303–315. [CrossRef] [PubMed]

7.   Kumar, N.; Pruthi, V. Potential applications of ferulic acid from natural sources. *Biotechnol. Rep.* **2014**, *4*, 86–93. [CrossRef] [PubMed]

8.   Davis, C.D.; Milner, J.A. Gastrointestinal microflora, food components and colon cancer prevention. *J. Nutr. Biochem.* **2009**, *20*, 743–752. [CrossRef] [PubMed]

9.   Wang, J.; Cao, Y.; Sun, B.; Wang, C. Characterisation of inclusion complex of trans-ferulic acid and hydroxypropyl-β-cyclodextrin. *Food Chem.* **2011**, *124*, 1069–1075. [CrossRef]

10.   Aceituno-Medina, M.; Mendoza, S.; Rodríguez, B.A.; Lagaron, J.M.; López-Rubio, A. Improved antioxidant capacity of quercetin and ferulic acid during in-vitro digestion through encapsulation within food-grade electrospun fibers. *J. Funct. Foods* **2015**, *12*, 332–341. [CrossRef]

11.   Celebioglu, A.; Uyar, T. Cyclodextrin nanofibers by electrospinning. *Chem. Commun.* **2010**, *46*, 6903–6905. [CrossRef] [PubMed]

12.   Celebioglu, A.; Uyar, T. Electrospinning of polymer-free nanofibers from cyclodextrin inclusion complexes. *Langmuir* **2011**, *27*, 6218–6226. [CrossRef] [PubMed]

13.   Zhao, R.; Wang, Y.; Li, X.; Sun, B.; Jiang, Z.; Wang, C. Water-insoluble sericin/β-cyclodextrin/PVA composite electrospun nanofibers as effective adsorbents towards methylene blue. *Colloids Surf. B* **2015**, *136*, 375–382. [CrossRef] [PubMed]

14.   Aytac, Z.; Uyar, T. Antioxidant activity and photostability of α-tocopherol/β-cyclodextrin inclusion complex encapsulated electrospun polycaprolactone nanofibers. *Eur. Polym. J.* **2016**, *79*, 140–149. [CrossRef]

15.   Kayaci, F.; Ertas, Y.; Uyar, T. Enhanced thermal stability of eugenol by cyclodextrin inclusion complex encapsulated in electrospun polymeric nanofibers. *J. Agric. Food. Chem.* **2013**, *61*, 8156–8165. [CrossRef] [PubMed]

16.   Mendes, A.C.; Stephansen, K.; Chronakis, I.S. Electrospinning of food proteins and polysaccharides. *Food Hydrocoll.* **2017**, *68*, 53–68. [CrossRef]

17.   Bhushani, J.A.; Anandharamakrishnan, C. Electrospinning and electrospraying techniques: Potential food based applications. *Trends Food Sci. Technol.* **2014**, *38*, 21–33. [CrossRef]

18.   Rezaei, A.; Nasirpour, A.; Fathi, M. Application of cellulosic nanofibers in food science using electrospinning and its potential risk. *Compr. Rev. Food Sci. Food Saf.* **2015**, *14*, 269–284. [CrossRef]

19.   Ghorani, B.; Tucker, N. Fundamentals of electrospinning as a novel delivery vehicle for bioactive compounds in food nanotechnology. *Food Hydrocoll.* **2015**, *51*, 227–240. [CrossRef]

20.   Celebioglu, A.; Uyar, T.J.J.; Chemistry, F. Antioxidant vitamin E/cyclodextrin inclusion complex electrospun nanofibers: Enhanced water solubility, prolonged shelf life, and photostability of vitamin E. *J. Agric. Food. Chem.* **2017**, *65*, 5404–5412. [CrossRef] [PubMed]

21.   Li, W.; Zhang, C.; Chi, H.; Li, L.; Lan, T.; Han, P.; Chen, H.; Qin, Y. Development of antimicrobial packaging film made from poly (lactic acid) incorporating titanium dioxide and silver nanoparticles. *Molecules* **2017**, *22*, 1170. [CrossRef] [PubMed]

22.   Makaremi, M.; Pasbakhsh, P.; Cavallaro, G.; Lazzara, G.; Aw, Y.K.; Lee, S.M.; Milioto, S. Effect of morphology and size of halloysite nanotubes on functional pectin bionanocomposites for food packaging applications. *ACS Appl. Mater. Interfaces* **2017**, *9*, 17476–17488. [CrossRef] [PubMed]

23.   Bertolino, V.; Cavallaro, G.; Lazzara, G.; Milioto, S.; Parisi, F. Halloysite nanotubes sandwiched between chitosan layers: Novel bionanocomposites with multilayer structures. *New J. Chem.* **2018**, *42*, 8384–8390. [CrossRef]

24.   Ago, M.; Okajima, K.; Jakes, J.E.; Park, S.; Rojas, O.J. Lignin-based electrospun nanofibers reinforced with cellulose nanocrystals. *Biomacromolecules* **2012**, *13*, 918–926. [CrossRef] [PubMed]

25. Wang, Y.; Chen, L. Fabrication and characterization of novel assembled prolamin protein nanofabrics with improved stability, mechanical property and release profiles. *J. Mater. Chem.* **2012**, *22*, 21592–21601. [CrossRef]

26. Wang, X.; Yu, D.-G.; Li, X.-Y.; Bligh, S.A.; Williams, G.R. Electrospun medicated shellac nanofibers for colon-targeted drug delivery. *Int. J. Pharm.* **2015**, *490*, 384–390. [CrossRef] [PubMed]

27. Malekzad, H.; Mirshekari, H.; Sahandi Zangabad, P.; Moosavi Basri, S.; Baniasadi, F.; Sharifi Aghdam, M.; Karimi, M.; Hamblin, M.R. Plant protein-based hydrophobic fine and ultrafine carrier particles in drug delivery systems. *Crit. Rev. Biotechnol.* **2018**, *38*, 47–67. [CrossRef] [PubMed]

28. Wu, W.; Kong, X.; Zhang, C.; Hua, Y.; Chen, Y. Improving the stability of wheat gliadin nanoparticles–Effect of gum arabic addition. *Food Hydrocoll.* **2018**, *80*, 78–87. [CrossRef]

29. Tarhini, M.; Greige-Gerges, H.; Elaissari, A. Protein-based nanoparticles: From preparation to encapsulation of active molecules. *Int. J. Pharm.* **2017**, *522*, 172–197. [CrossRef] [PubMed]

30. Kfoury, M.; Auezova, L.; Greige-Gerges, H.; Ruellan, S.; Fourmentin, S.J.F.C. Cyclodextrin, an efficient tool for trans-anethole encapsulation: Chromatographic, spectroscopic, thermal and structural studies. *Food Chem.* **2014**, *164*, 454–461. [CrossRef] [PubMed]

31. Hong, N.V.; Trujillo, E.; Puttemans, F.; Jansens, K.J.; Goderis, B.; Van Puyvelde, P.; Verpoest, I.; Van Vuure, A.W. Developing rigid gliadin based biocomposites with high mechanical performance. *Compos. Part A* **2016**, *85*, 76–83. [CrossRef]

32. Yang, H.; Feng, K.; Wen, P.; Zong, M.-H.; Lou, W.-Y.; Wu, H. Enhancing oxidative stability of encapsulated fish oil by incorporation of ferulic acid into electrospun zein mat. *Food Sci. Technol.* **2017**, *84*, 82–90. [CrossRef]

33. Aytac, Z.; Ipek, S.; Durgun, E.; Uyar, T.J.J. Antioxidant electrospun zein nanofibrous web encapsulating quercetin/cyclodextrin inclusion complex. *J. Mater. Sci.* **2018**, *53*, 1527–1539. [CrossRef]

34. European Commission. Commission Regulation (EU) No 10/2011 of 14 January 2011 on plastic materials and articles intended to come into contact with food. *Off. J. Eur. Union* **2011**, *12*, 1–89.

35. Atay, E.; Fabra, M.J.; Martínez-Sanz, M.; Gomez-Mascaraque, L.G.; Altan, A.; Lopez-Rubio, A.J.F.H. Development and characterization of chitosan/gelatin electrosprayed microparticles as food grade delivery vehicles for anthocyanin extracts. *Food Hydrocoll.* **2018**, *77*, 699–710. [CrossRef]

36. Zhang, W.; Li, X.; Yu, T.; Yuan, L.; Rao, G.; Li, D.; Mu, C.J.F.R.I. Preparation, physicochemical characterization and release behavior of the inclusion complex of trans-anethole and β-cyclodextrin. *Food Res. Int.* **2015**, *74*, 55–62. [CrossRef] [PubMed]

37. Olga, G.; Styliani, C.; Ioannis, R.G. Coencapsulation of ferulic and gallic acid in hp-b-cyclodextrin. *Food Chem.* **2015**, *185*, 33–40. [CrossRef] [PubMed]

38. Aytac, Z.; Kusku, S.I.; Durgun, E.; Uyar, T.J.M.S. Encapsulation of gallic acid/cyclodextrin inclusion complex in electrospun polylactic acid nanofibers: Release behavior and antioxidant activity of gallic acid. *Mater. Sci. Eng. C* **2016**, *63*, 231–239. [CrossRef] [PubMed]

39. Hu, L.; Zhang, H.; Song, W.; Gu, D.; Hu, Q.J.C. Investigation of inclusion complex of cilnidipine with hydroxypropyl-β-cyclodextrin. *Carbohydr. Polym.* **2012**, *90*, 1719–1724. [CrossRef] [PubMed]

40. Celebioglu, A.; Umu, O.C.; Tekinay, T.; Uyar, T. Antibacterial electrospun nanofibers from triclosan/cyclodextrin inclusion complexes. *Colloids Surf. B* **2014**, *116*, 612–619. [CrossRef] [PubMed]

41. Liu, B.; Zeng, J.; Chen, C.; Liu, Y.; Ma, H.; Mo, H.; Liang, G.J.F. Interaction of cinnamic acid derivatives with β-cyclodextrin in water: Experimental and molecular modeling studies. *Food Chem.* **2016**, *194*, 1156–1163. [CrossRef] [PubMed]

42. Ram, M.; Seitz, L.M.; Dowell, F.E. Natural fluorescence of red and white wheat kernels. *Cereal Chem.* **2004**, *81*, 244–248. [CrossRef]

43. Liu, B.; Zhu, X.; Zeng, J.; Zhao, J. Preparation and physicochemical characterization of the supramolecular inclusion complex of naringin dihydrochalcone and hydroxypropyl-β-cyclodextrin. *Food Res. Int.* **2013**, *54*, 691–696. [CrossRef]

44. Sharif, N.; Golmakani, M.-T.; Niakousari, M.; Ghorani, B.; Lopez-Rubio, A. Food-grade gliadin microstructures obtained by electrohydrodynamic processing. *Food Res. Int.* **2018**. [CrossRef]

45. Uyar, T.; Besenbacher, F.J.P. Electrospinning of uniform polystyrene fibers: The effect of solvent conductivity. *Polymer* **2008**, *49*, 5336–5343. [CrossRef]

46. Wendorff, J.H.; Agarwal, S.; Greiner, A. *Electrospinning: Materials, Processing, and Applications*; John Wiley & Sons: Hoboken, NJ, USA, 2012; p. 242.

47. Ramakrishna, S.; Fujihara, K.; Teo, W.E.; Lim, T.C.; Ma, Z. *An Introduction to Electrospinning and Nanofibers*; World Scientific Publishing Company: Singapore, 2005; pp. 90–103. ISBN 981-256-415-2.

48. Granata, G.; Consoli, G.M.; Nigro, R.L.; Geraci, C.J.F. Hydroxycinnamic acids loaded in lipid-core nanocapsules. *Food Chem.* **2018**, *245*, 551–556. [CrossRef] [PubMed]

49. Aceituno-Medina, M.; Lopez-Rubio, A.; Mendoza, S.; Lagaron, J.M. Development of novel ultrathin structures based in amaranth (*Amaranthus hypochondriacus*) protein isolate through electrospinning. *Food Hydrocoll.* **2013**, *31*, 289–298. [CrossRef]

50. Torres-Giner, S.; Gimenez, E.; Lagarón, J.M. Characterization of the morphology and thermal properties of zein prolamine nanostructures obtained by electrospinning. *Food Hydrocoll.* **2008**, *22*, 601–614. [CrossRef]

51. Yang, J.-M.; Zha, L.; Yu, D.-G.; Liu, J. Coaxial electrospinning with acetic acid for preparing ferulic acid/zein composite fibers with improved drug release profiles. *Colloids Surf. B* **2013**, *102*, 737–743. [CrossRef] [PubMed]

52. Wang, H.; Hao, L.; Niu, B.; Jiang, S.; Cheng, J.; Jiang, S. Kinetics and antioxidant capacity of proanthocyanidins encapsulated in zein electrospun fibers by cyclic voltammetry. *J. Agric. Food. Chem.* **2016**, *64*, 3083–3090. [CrossRef] [PubMed]

53. Yoksan, R.; Jirawutthiwongchai, J.; Arpo, K.J.C.; Biointerfaces, S.B. Encapsulation of ascorbyl palmitate in chitosan nanoparticles by oil-in-water emulsion and ionic gelation processes. *Colloids Surf. B* **2010**, *76*, 292–297. [CrossRef] [PubMed]

54. Fiddler, W.; Parker, W.E.; Wasserman, A.E.; Doerr, R.C. Thermal decomposition of ferulic acid. *J. Agric. Food Chem.* **1967**, *15*, 757–761. [CrossRef]

55. Neo, Y.P.; Ray, S.; Jin, J.; Gizdavic-Nikolaidis, M.; Nieuwoudt, M.K.; Liu, D.; Quek, S.Y. Encapsulation of food grade antioxidant in natural biopolymer by electrospinning technique: A physicochemical study based on zein–gallic acid system. *Food Chem.* **2013**, *136*, 1013–1021. [CrossRef] [PubMed]

56. Kayaci, F.; Sen, H.S.; Durgun, E.; Uyar, T.J.F. Functional electrospun polymeric nanofibers incorporating geraniol–cyclodextrin inclusion complexes: High thermal stability and enhanced durability of geraniol. *Food Res. Int.* **2014**, *62*, 424–431. [CrossRef]

57. Hartley, R.; Jones, E.J.J. Effect of ultraviolet light on substituted cinnamic acids and the estimation of their cis and trans isomers by gas chromatography. *J. Chromatogr. A* **1975**, *107*, 213–218. [CrossRef]

58. Del Valle, E.M.J.P. Cyclodextrins and their uses: A review. *Process Biochem.* **2004**, *39*, 1033–1046. [CrossRef]

59. Anselmi, C.; Centini, M.; Maggiore, M.; Gaggelli, N.; Andreassi, M.; Buonocore, A.; Beretta, G.; Facino, R.M. Non-covalent inclusion of ferulic acid with α-cyclodextrin improves photo-stability and delivery: NMR and modeling studies. *J. Pharm. Biomed. Anal.* **2008**, *46*, 645–652. [CrossRef] [PubMed]

60. Bibby, D.C.; Davies, N.M.; Tucker, I.G. Mechanisms by which cyclodextrins modify drug release from polymeric drug delivery systems. *Int. J. Pharm.* **2000**, *197*, 1–11. [CrossRef]

© 2018 by the authors. Licensee MDPI, Basel, Switzerland. This article is an open access article distributed under the terms and conditions of the Creative Commons Attribution (CC BY) license (http://creativecommons.org/licenses/by/4.0/).

nanomaterials

MDPI

Article

Anchoring Gated Mesoporous Silica Particles to Ethylene Vinyl Alcohol Films for Smart Packaging Applications

Virginia Muriel-Galet [1], Édgar Pérez-Esteve [2], María Ruiz-Rico [2], Ramón Martínez-Máñez [3,4], José Manuel Barat [2], Pilar Hernández-Muñoz [1] and Rafael Gavara [1,*]

[1] Instituto de Agroquímica y Tecnología de Alimentos, IATA-CSIC, Grupo de Envases, Av. Agustin Escardino 7, 46980 Paterna, Spain; vmurielgalet@gmail.com (V.M.-G.); phernan@iata.csic.es (P.H.-M.)
[2] Departamento de Tecnología de Alimentos, Grupo de Investigación e Innovación Alimentaria, Universitat Politècnica de València. Camino de Vera s/n, 46022 Valencia, Spain; edpees@upv.es (É.P.-E.); maruiri@etsia.upv.es (M.R.-R.); jmbarat@tal.upv.es (J.M.B.)
[3] Instituto Interuniversitario de Investigación de Reconocimiento Molecular y Desarrollo Tecnológico (IDM), Universitat Politècnica de València, Universitat de València. Departamento de Química, Universitat Politècnica de València, Camino de Vera s/n, 46022 Valencia, Spain; rmaez@qim.upv.es
[4] CIBER de Bioingeniería, Biomateriales y Nanomedicina (CIBER-BBN), Camino de Vera s/n, 46022 Valencia, Spain
* Correspondence: rgavara@iata.csic.es; Tel.: +34-963-900-022

Received: 2 October 2018; Accepted: 20 October 2018; Published: 22 October 2018

Abstract: This work is a proof of concept for the design of active packaging materials based on the anchorage of gated mesoporous silica particles with a pH triggering mechanism to a packaging film surface. Mesoporous silica micro- and nanoparticles were loaded with rhodamine B and functionalized with N-(3-trimethoxysilylpropyl)diethylenetriamine. This simple system allows regulation of cargo delivery as a function of the pH of the environment. In parallel, poly(ethylene-*co*-vinyl alcohol) films, EVOH 32 and EVOH 44, were ultraviolet (UV) irradiated to convert hydroxyl moieties of the polymer chains into –COOH functional groups. The highest COOH surface concentration was obtained for EVOH 32 after 15 min of UV irradiation. Anchoring of the gated mesoporous particles to the films was carried out successfully at pH 3 and pH 5. Mesoporous particles were distributed homogeneously throughout the film surface and in greater concentration for the EVOH 32 films. Films with the anchored particles were exposed to two liquid media simulating acidic food and neutral food. The films released the cargo at neutral pH but kept the dye locked at acidic pH. The best results were obtained for EVOH 32 irradiated for 15 min, treated for particle attachment at pH 3, and with mesoporous silica nanoparticles. This opens the possibility of designing active materials loaded with antimicrobials, antioxidants, or aromatic compounds, which are released when the pH of the product approaches neutrality, as occurs, for instance, with the release of biogenic amines from fresh food products.

Keywords: MCM-41; gated mesoporous silica particles; EVOH films; anchorage on film surface; active packaging; pH-mediated delivery

---

1. Introduction

Active packaging is a novel technology in which the packaging system is designed to actively improve the stability and/or quality of the packaged product from processing to consumption. This technology is being implemented in various industries, although, owing to the fast perishability of food, pharmaceutical, and cosmetic products, these are the areas that receive the most attention [1,2]. The mechanism of action is basically related to the release or retention of substances whose presence or absence is important for the product's stability or quality. By suitably combining design and mechanism of action, various active packaging technologies were created, including oxygen scavengers, ethylene scavengers, humidity controllers, aroma releasers, antioxidant or antimicrobial releasers, enzyme-based systems, etc. Several interesting reviews were published on this subject [3,4].

There are basically two procedures for designing active packaging systems: manufacture of an independent device that contains the active agent [5], or manufacture of active materials that incorporate the agent in the package wall or on the wall surface [6]. Of these two general procedures, the manufacture of active packaging materials is gaining attention over the development of independent devices. In the latter, the presence of a device often labeled as toxic because it contains an inedible component in contact with the product is not well accepted by consumers. Thus, the incorporation of active agents on or in the polymeric films that constitute the package walls is the preferred option. Owing to their at least partially amorphous morphology and the presence of free volume (voids) in polymer matrices, polymeric films allow mass transport (permeation, migration, scalping, or sorption), processes that are profitable for the design of active materials [6]. However, some precautions have to be considered in these designs. Firstly, the incorporation of the active substance in the package walls or the action of the substance should not modify the functional properties of the packaging throughout its use. Secondly, the substance should not lose activity owing to interactions or degradations caused by the film manufacturing procedure. Thirdly, the mechanism of action (release, adsorption) should be maintained and controlled. Finally, and most importantly, the packaging system should include a triggering mechanism to avoid premature action and partial exhaustion of the system prior to the presence of the product to be protected. This last condition was achieved through various procedures: humidity-activated systems [7,8], temperature-activated systems [9], or radiation-activated systems [10]. Another approach focuses on including substances covalently anchored to the package walls that exert their action via direct contact with the packaged product, that is, no agent is released or captured. Such systems were successfully prepared, for instance, via oxidation of a conventional film surface and anchorage of enzymes [11,12], antimicrobials [13] (such as lysozyme), or antioxidants [14]. Indeed, Goddard, Talbert and Hotchkiss [11] successfully functionalized a polyethylene surface with lactase to design an active package that reduced the amount of lactose in milk. Muriel Galet, Talbert, Hernandez Munoz, Gavara and Goddard [12] anchored lysozymes to the surface of ethylene vinyl alcohol (EVOH) to generate a film with antimicrobial properties against *Listeria monocytogenes*. Similarly, Saini, Sillard, Belgacem and Bras [13] anchored a bacteriocin to cellulose fibers with potential application in active food packaging. Roman, Decker and Goddard [14] prepared an antioxidant active film via functionalization of a polypropylene surface with polyphenols generated by the action of laccase. Vasile, et al. [15] covalently bonded chitosan to plasma-treated polyethylene and obtained a material with antimicrobial properties against *Salmonella enteriditis*, *Escherichia coli*, and *Listeria monocytogenes*.

Considered from another point of view, new technologies based on nanomaterials or nanocomposites received massive attention in the packaging field lately, especially in active packaging research. The new properties and functions of nanoscale particles display new opportunities for enhancing traditional product performance. A wide range of nanostructured materials were included as fillers in packaging films to provide barrier properties or improved mechanical resistance, or to control activity in smart packaging applications [16,17]. Moreover, nanostructures were also included to provide antioxidant and antimicrobial properties. Biddeci, et al. [18] reported the design of a pectin-based biopolymer film with both antioxidant and antimicrobial activities. The film was created by filling the pectin matrix with modified halloysite nanotubes containing essential peppermint oil. Later, Li, et al. [19] used the solvent volatilization method to prepare polylactide films containing nanoparticles of silver and titanium dioxide. The developed films showed good antimicrobial activity against two common food pathogens: *E. coli* and *Listeria monocytogenes*.

Among commented nanostructured materials, mesoporous silica particles (MSPs) exhibit unique features such as high stability, biocompatibility, nontoxicity, and large load capacity. Moreover, the possibility of functionalizing the external surface with gate-like ensembles makes these materials unique candidates for the design of on-command controlled release devices. In fact, a number of gated materials based on mesoporous silica particles able to deliver the cargo upon application of target physical (such as light or temperature) [20,21], chemical (pH changes or redox potential) [22,23], and biochemical (enzymes, antibodies, or DNA) [24] stimuli were reported. These functionalized MSPs are mainly used in the fields of drug delivery [25–28] or sensing [29], and are prepared in the form of nanoparticles [30] or microparticles [31]. In contrast, gated MSPs are barely incorporated in polymers or on surfaces. Moreover, as far as we are aware, MSPs were never previously used in the design of active packaging.

In this scenario, we report herein the preparation of smart films based on poly(ethylene-*co*-vinyl alcohol) (EVOH)-containing gated MSPs covalently anchored and able to modulate the release of a model molecule in response to changes in the pH of the environment. EVOHs are a family of copolymers with different ethylene molar percentages, commonly used in packaging technologies as they provide an excellent oxygen barrier thanks to their high crystallinity ratio and the high cohesive energy density caused by the large number of hydrogen bonds between macromolecular chains. Moreover, EVOH not only provides hydrophilicity, but also contains suitable hydroxyl sites for functionalization [12].

In most reported works about the inclusion of micro- or nanoparticles in packaging systems, functional nanofillers are mixed with the polymer via three procedures: the solvent casting method, the melt mixing method, and in situ polymerization [32]. However, these procedures are not suitable for the present design because the MSP particles require to be exposed to conditions that would open the gates, promoting the release of the agent included in the particle during preparation. Moreover, if the particle was included in the polymer matrix, gate opening would be impeded or delayed.

2. Materials and Methods

2.1. Chemicals and Reagents

Tetraethylorthosilicate (TEOS), *N*-cetyltrimethylammonium bromide (CTABr), triethanolamine (TEAH3), sodium hydroxide (NaOH), acetonitrile, and *N*-(3-trimethoxysilylpropyl)diethylenetriamine (N3) were provided by Sigma (Sigma-Aldrich Química S.L., Madrid, Spain). Films, 75-μm-thick, of Soarnol DC3203FB ethylene vinyl alcohol copolymer with 32% ethylene molar content (EVOH 32) and Soarnol AT4403B with 44% ethylene molar content (EVOH 44) were kindly provided by The Nippon Synthetic Chemical Company (Osaka, Japan). Isopropanol, acetone, acetic acid, 1-ethyl-3-(3-dimethylaminopropyl) carbodiimide (EDC), *N*-hydroxysuccinimide (NHS), and Toluidine Blue O (TBO) were purchased from Sigma (Sigma-Aldrich, Madrid, Spain). Water was obtained from a Milli-Q Plus purification system (Millipore, Molsheim, France).

2.2. Synthesis of Mesoporous Silica Particless

Microparticulated MCM-41 particles (**M**) were synthesized following the so-called "atrane route", according to the method described by Perez-Esteve, et al. [33]. *N*-cetyltrimethylammonium bromide (acting as a structure-directing agent) was added to a solution of triethanolamine (TEAH3) containing sodium hydroxide (NaOH) and tetraethylorthosilicate (TEOS). Temperature was then set at 118 °C. After the CTABr was dissolved in the solution, water was slowly added with vigorous stirring at 70 °C. This mixture was aged in an autoclave at 100 °C for 24 h. The molar ratio of the reagents was fixed at 7 TEAH3:2 TEOS:0.52 CTABr:0.5 NaOH:180 H_2O.

Nanoparticulated MCM-41 particles (**N**) were synthesized using the procedure described by Perez-Esteve, Fuentes, Coll, Acosta, Bernardos, Amoros, Marcos, Sancenon, Martinez-Manez and Barat [33]. *N*-cetyltrimethylammonium bromide was firstly dissolved in 480 mL of deionized water. Then, 3.5 mL of a sodium hydroxide solution was then added, and the mixture was heated to 80 °C.

Finally, TEOS was added dropwise to the surfactant solution. The mixture was stirred for 2 h to give a white precipitate. The molar ratio of the reagents was fixed at 1 TEOS:0.1 CTABr:0.27 NaOH:1000 H_2O.

After synthesis, the resulting microparticulated or nanoparticulated powder was recovered by centrifugation, washed with deionized water, and air-dried at room temperature. To prepare the final mesoporous materials, the as-synthesized solids were calcined at 550 °C using an oxidant atmosphere for 5 h in order to remove the template phase.

2.3. Mesoporous Silica Particle Loading and Functionalization

Once the starting supports (**M** and **N**) were synthesized, both supports were loaded with rhodamine B (**M-Rh** and **N-Rh**). Amounts of 100 mg of template-free MCM-41 and 39 mg of rhodamine B dye (0.8 mmol Rhodamine B/g MCM-41) were suspended in 25 mL of acetonitrile inside a round-bottom flask in an inert atmosphere. The mixture was then stirred for 24 h at room temperature to achieve maximum loading in the MCM-41 scaffolding pores.

To obtain loaded and functionalized solids (**M-Rh-N3** and **N-Rh-N3**), an excess of N3 (0.43 mL, 0.015 mmol) was added to the mixtures. The final mixtures were stirred for 5.5 h at room temperature. The two loaded and functionalized solids were then isolated by vacuum filtration, washed with 300 mL of water adjusted to pH 2, and dried at room temperature for 24 h.

2.4. Characterization of Mesoporous Silica Particles

Mesoporous silica particles were characterized by means of powder X-ray diffraction (PXRD), N_2 adsorption–desorption isotherms, zeta potential, thermogravimetric analyses, and microscopy.

PXRD was performed on a D8 Advance diffractometer (Bruker, Coventry, UK) using CuKα radiation. N_2 adsorption–desorption isotherms were recorded with a Micromeritics ASAP 2010 automated sorption analyzer (Micromeritics Instrument Corporation, Norcross, GA, USA). The samples were degassed at 120 °C in vacuum overnight. The specific surface areas were calculated from the adsorption data in the low pressure range using the Brunauer–Emmett–Teller (BET) model. Pore size was determined following the Barrett–Joyner–Halenda (BJH) method. From the XRD and porosimetry studies, the a_0 cell parameter and wall thickness of the various supports were calculated.

The functionalization degree of different particles was estimated by determining the percentage of organic matter in functionalized particles and confirmed by zeta potential measurements. The percentage of organic matter was determined by thermogravimetric analyses (TGA) on a TGA/SDTA 851e Mettler Toledo balance, using an oxidant atmosphere (air, 80 mL/min) with a heating program consisting of a heating ramp of 10 °C per minute from 393 to 1273 K and an isothermal heating step at this temperature for 30 min. The percentage of lost matter in the 100–750 °C range was used to estimate the functionalization degree since the only organic matter in the particle was due to the anchored amines. To determine the zeta potential (α) of bare and functionalized MSP, a Zetasizer Nano ZS unit (Malvern Instruments, Malvern, UK) was used. Samples were dispersed in distilled water at a concentration of 1 mg/mL. Before each measurement, samples were sonicated for 2 min to preclude aggregation, and the particle dispersions were carefully placed in a folded capillary zeta cell (Malvern Instruments, Malvern, UK). The zeta potential was calculated from the particle mobility values by applying the Smoluchowski model. The average of five recordings is reported as the zeta potential. The measurement was performed at 25 °C. Measurements were performed in triplicate.

For transmission electron microscopy (TEM) analysis, MSPs were dispersed in dichloromethane and sonicated for 2 min to preclude aggregates, and the suspension was deposited onto copper grids coated with carbon film (Aname SL, Madrid, Spain). Imaging of the MSP samples was performed using a JEOL JEM-1010 (JEOL Europe SAS, Croissy, France) operating at an acceleration voltage of 80 kV. Field-emission scanning electron microscopy (FESEM) images were acquired with a Zeiss Ultra 55 (Carl Zeiss NTS GmbH, Oberkochen, Germany) operating at 1.5 mV and a working distance of 5.6 mm. Observations were done in the secondary electron mode.

2.5. Preparation and Functionalization of EVOH Films

EVOH-32 and EVOH-44 films were cut into 4-cm^2 pieces and were sequentially cleaned in iso-propanol, acetone, and deionized water. The EVOH films were sonicated twice with each solvent at 10 min intervals. Clean films were left to dry in Petri dishes inside a desiccator with anhydrous calcium sulfate at room temperature (25 °C) for 12 h.

EVOH-32 and EVOH-44 films were irradiated in open glass Petri dishes for 1, 3, 10, and 15 min under a vacuum ultraviolet (UV) Xe excimer lamp with 6 W at 172 nm (UV-Consulting Peschl España S.L., Valencia, Spain). The films were turned over and exposed to UV light under the same conditions. This method was used to oxidize and create carboxylic acid functional groups on both film surfaces.

In order to select the more suitable EVOH film (EVOH-32 or EVOH-44) for the subsequent attachment of N3-MSP, a quantitative method for determining the number of carboxylic acids created after UV irradiation, the Toluidine Blue O (TBO) dye assay, was carried out. The method used was that described by Hernandez, Tseng, Wong, Stoddart and Zink [22], Uchida, et al. [34], and Kang, et al. [35] with some modifications. In brief, control and UV-treatment films were immersed in a TBO solution (0.5 mM TBO solution in deionized water with the pH adjusted to 10.0 with 0.5 M NaOH) and shaken for 2 h at room temperature (25 °C). Then, the films were rinsed three times with deionized water adjusted to pH 10.0 to remove non-complexed dye. To desorb the complexed dye on the film surfaces, films were submerged in 50 wt% acetic acid solution for 15 min. The absorbance of the acetic acid solutions was measured at 633 nm using a UV–visible light (UV–Vis) spectrophotometer (Agilent 8453 Spectroscopy System), and compared with a standard curve of TBO dye in 50 wt% acetic acid solution.

2.6. N3-MSP Deposition

To attach **M-Rh-N3** and **N-Rh-N3** to the EVOH films via covalent bonding, a previously reported procedure was used [12]. EVOH films were stirred for 30 min in two conjugation buffers containing 5×10^{-2} M EDC and 5×10^{-3} M NHS at pH 3.0 or 5.0 to select the most adequate bonding conditions. The selected concentrations represent molar excesses of at least 10-fold and 100-fold for EDC and NHS, respectively, compared to the determined mole quantity of surface carboxylic acid groups. Then EVOH films (4 cm^2/mL) were sonicated for 30 min in buffer (pH 3.0 or pH 5.0) containing **M-Rh-N3** or **N-Rh-N3** at a final concentration of 0.5 mg/mL, and then stirred for 2 h at room temperature (24 ± 1 °C). **EVOH-M-Rh-N3** and **EVOH-N-Rh-N3** films were rinsed with buffer (at pH 3.0 or pH 5.0) to clean unattached particles, before being dried, and stored in dry conditions until use. The whole process is summarized in Figure 1.

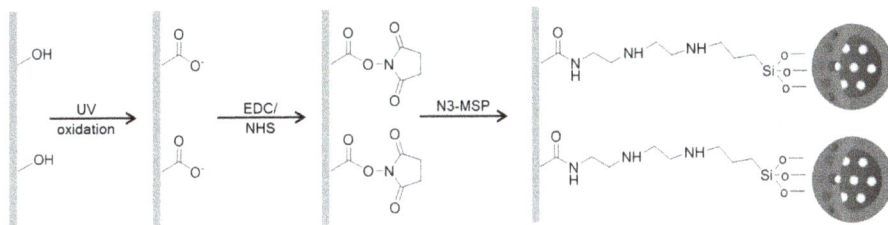

Figure 1. Scheme of the functionalization process: surface modification of ethylene vinyl alcohol (EVOH) and the subsequent carbodiimide-mediated anchoring of N-(3-trimethoxysilylpropyl) diethylenetriamine (N3)-functionalized mesoporous silica particles (MSPs).

2.7. Surface Analysis

The efficiency of immobilization of N3-MSP on the EVOH films was studied by means of FE-SEM. FESEM images were acquired with a Zeiss Ultra 55 (Carl Zeiss NTS GmbH, Oberkochen, Germany) and observed in the secondary electron mode. Micrographs of the particles before and after immobilization on the EVOH films were obtained.

2.8. Controlled Release from the Films

In a typical experiment, 1 cm^2 of film was suspended in 4 mL of water adjusted to pH 2 and pH 7.5. At certain times (2 min, 1, 2, 4, 6, 8, and 24 h), aliquots were separated and filtered. Dye released from the pore voids to the aqueous solution was quantified by measuring the emission band of rhodamine B centered at 580 nm (excitation at 554 nm) using a Jasco-FP-8500 spectrofluorometer (Tokyo, Japan).

The rhodamine B release kinetics from pore voids of the porous silica supports were calculated using the Higuchi model, where the amount of guest release, Q_t, per unit of exposed area at time t can then be described by the following equation:

$$Q_t = k_H \times \sqrt{t},$$

where k_H is the release rate constant for the Higuchi model.

3. Results and Discussion

3.1. MSP Preparation and Characterization

Microparticulated (**M**) and nanoparticulated (**N**) MSPs as synthesized, MSPs loaded with rhodamine B (**M-Rh and N-Rh**), and MSPs functionalized with *N*-(3-trimethoxysilylpropyl) diethylenetriamine (**N3**) (**M-Rh-N3** for micro and **N-Rh-N3** for nano) were prepared and characterized using standard procedures. Figure 2a shows powder X-ray patterns of the MCM-41 microparticles as synthesized, after calcination, loaded once with Rhodamine B and functionalized with N3 (**M-Rh-N3**).

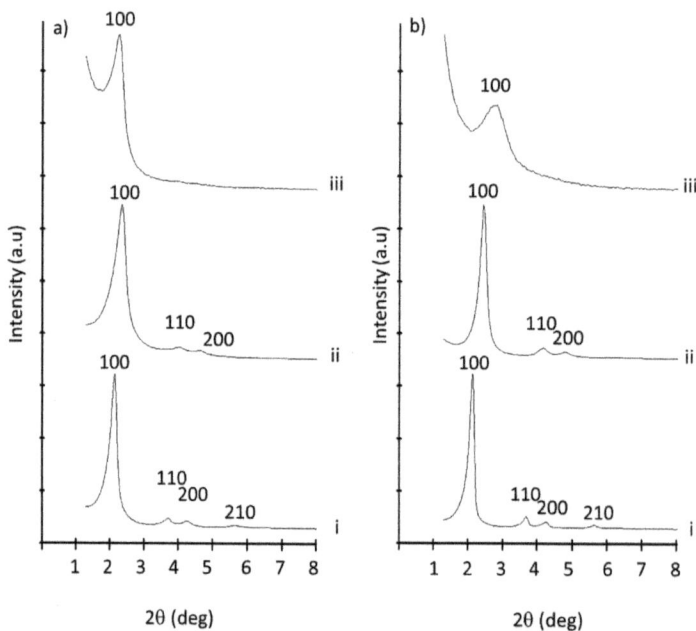

Figure 2. Powder X-ray diffraction of MCM-41 particles as prepared (i), after calcination (ii), and after the loading and functionalization process: (**a**) microparticles; (**b**) nanoparticles.

The PXRD of the microparticulated MSPs as synthesized (Figure 2a, curve i) shows four low-angle reflections typical of a hexagonal array that can be indexed as (100), (110), (200), and (210) Bragg peaks. A significant displacement of the (100) peak in the diffractogram was clearly observed for the calcined microparticles (Figure 2a, curve ii), corresponding to a cell contraction of ca. 4 Å. This displacement and the broadening of the (110) and (200) peaks are most likely related to further condensation of silanol groups during the calcination step. Figure 2a, curve iii corresponds to the **M-Rh-N3** PXRD pattern. In this case, a slight intensity decrease and a further broadening of the (110) and (200) reflections were observed, probably due to a loss of contrast owing to the filling of the pore voids with the dye and the functionalization with amines. Nevertheless, the value and intensity of the (100) peak in this pattern clearly showed that both the loading process with the dye and the subsequent functionalization with amines did not damage the mesoporous scaffolding. The same diffractogram features were obtained for the solid materials prepared with MCM-41 nanoparticles (Figure 2b). Since both particles belong to the MCM-41 family, the similarity between diffractograms of micro- and nanoparticles was expected [31].

The preservation of the mesoporous structure in the final loaded and functionalized solids **M-Rh-N3** and **N-Rh-N3** was also confirmed by means of transmission electron microscopy (TEM). Figure 3 shows the different morphologies of the two types of particles. While MCM-41 microparticles (Figure 3a) are irregular particles with diameters ranging between 0.8 and 1.2 mm, MCM-41 nanoparticles (Figure 3c) show a spherical shape with diameters of ca. 100 nm. No significant differences were observed in particle size before and after functionalization. The images show the typical channels of the MCM-41 matrix both as alternate black and white stripes and as a pseudo-hexagonal array of pore voids in both types of particles. These channels are seen not only in the calcined material but also in the loaded and functionalized supports (Figure 3b,d), confirming that the initial morphology of the mesoporous matrix was maintained after the loading and functionalization process.

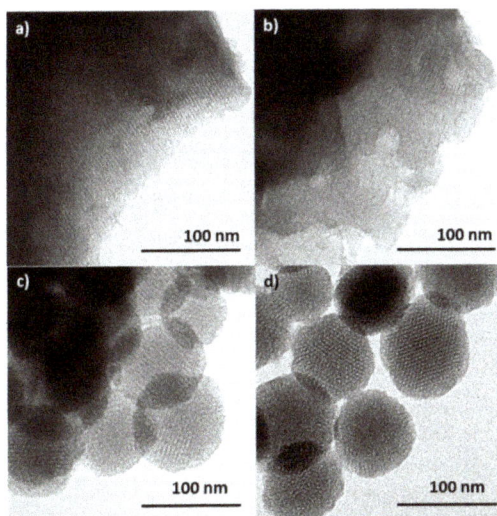

Figure 3. TEM images of calcined (**a**) microparticles (**M**), and (**c**) nanoparticles (**N**), showing the typical hexagonal porosity of the MCM-41 matrix. TEM images of solid MSP micro- and nanoparticles loaded with rhodamine B and functionalized with N3: (**b**) **M-Rh-N3**, and (**d**) **N-Rh-N3**.

Textural properties of the various supports calculated from the nitrogen adsorption–desorption isotherms are summarized in Table 1. As observed, both types of particles (micro and nano) presented similar textural properties (total area of ca. $1000\ m^2 \cdot g^{-1}$, pore volume of ca. $0.9\ c^3 \cdot g^{-1}$, and pore size of ca. 2.5 nm). These features were reported to be sufficient for encapsulation of molecules of special interest in food technology (i.e., antimicrobial agents, drugs, flavors, vitamins, antioxidants, enzymes, and other functional compounds) in MSPs [36]. Table 1 also shows that, after the loading and functionalization process, a decrease in the N_2 volume adsorbed was produced. This reduction is indicative of mesoporous systems with partially filled mesopores.

Table 1. Characterization of the mesoporous silica particles (MSPs) before and after functionalization: Brunauer–Emmett–Teller (BET) specific surface area, pore volume, and pore size calculated from the N_2 adsorption–desorption isotherms, content of rhodamine B and amines (α_{Rh} and α_{N3}, mg/g_{solid}), and zeta potential (Z-potential, mV). M—microparticles; N—nanoparticles; Rh—loaded with rhodamine B; N3—functionalized with N-(3-trimethoxysilylpropyl)diethylenetriamine.

Sample	BET Area (m^2/g)	Pore Volume (c^3/g)	Pore Size (nm)	α_{Rh} (mg/g_{solid})	α_{N3} (mg/g_{solid})	Z-Potential (mV)
M	1072	0.91	2.62	-	-	−38
M-Rh-N3	243	-	-	15.3	81	41
N	986	0.84	2.51	-	-	−36
N-Rh-N3	143	-	-	17.2	142	43

The content of organic matter in the final hybrid solids **M-Rh-N3** and **N-Rh-N3** was determined by thermogravimetric analysis. Contents (α) of rhodamine B and the amine derivative are shown in Table 1. The organic matter contents in both materials is similar to those reported by other authors for similar systems based on MSPs loaded with rhodamine B and functionalized with amines [37]. Table 1 also includes the zeta potential values of **M-Rh-N3** and **N-Rh-N3** suspended in distilled water. Bare micro- and nanoparticles showed negative zeta potential values of ca. −35 mV. These negative values are typical of bare mesoporous silica particles, which contain SiO– groups in their surfaces. After functionalization with amines, a neutralization of the silica by ammonium groups was produced, and zeta potential values changed from negative to positive values (ca. +40 mV). This inversion of the surface charge after organic functionalization was reported in literature for similar systems [38,39] and confirms the efficiency of the functionalization process.

3.2. EVOH Film Surface Analysis

Two EVOH copolymers with a different molar percentage of ethylene monomer (32%—EVOH 32, and 44%—EVOH 44) were selected as films for functionalization. The lower the ethylene content, the higher the hydrophilicity, the interchain interactions, the rigidity, and the gas barrier. The films were supplied as the central layer of a three-layer polypropylene (PP)/EVOH/PP coextruded film without tie layers, so the PP protector layers could be easily peeled off. Monolayer EVOH films were exposed to UV irradiation to oxidize the film surface. The presence of a significant amount of –OH substituents permits the generation of carboxylic substituents, which are required for **M-Rh-N3** and **N-Rh-N3** particle attachment via EDC/NHS.

The TBO assay was used to quantify the amount of carboxyl groups created after the irradiation on the surface of the EVOH 32 and EVOH 44 films (Figure 4) [40]. The non-zero initial relevant value for EVOH 32 is due to the presence of some carbonyl groups in the copolymer, as this family of materials is obtained via hydrolysis of statistical poly(ethylene-*co*-vinyl acetate) polymers.

As expected, the density of carboxylic groups increased significantly for both films with the length of UV treatment. For instance, the number of nmol of COOH/cm^2 for EVOH 32 films ranged from 23.31 nmol/cm^2 after 1 min to 75.49 nmol/cm^2 after 15 min of UV treatment. A similar trend was found for the EVOH 44 samples, although, in this case, the density of the generated carboxyl groups was significantly lower than that found for EVOH 32.

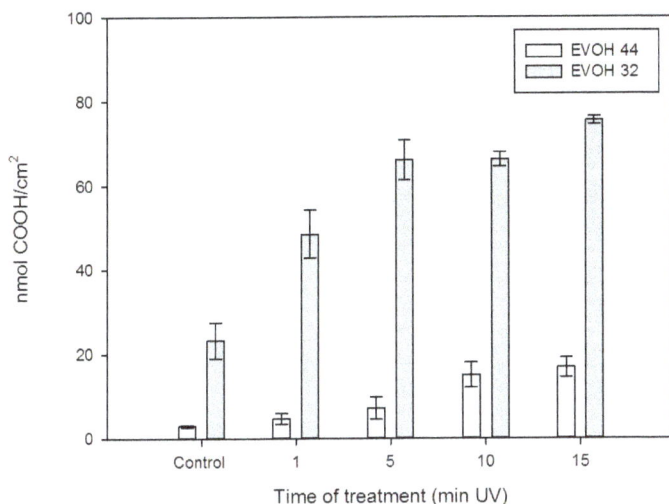

Figure 4. Surface density of COOH groups on EVOH 32 and EVOH 44 films calculated by the Toluidine Blue O (TBO) assay as a function of the ultraviolet (UV) treatment time.

3.3. MSP Immobilization

After confirming the activation of the EVOH films, the amine-functionalized mesoporous silica particles were covalently linked to the films through amide bonds in the presence of carbodiimide. The immobilization efficiency of **M-Rh-N3** and **N-Rh-N3** on the surface of the EVOH films was studied by means of FESEM. Micrographs of the gated MSPs before and after immobilization on the films are shown in Figure 5. Four films were developed, **EVOH 32-N-Rh-N3** in which the nanoparticles were anchored at pH = 5 (**FN5**) and at pH = 3 (**FN3**) (Figure 6), and **EVOH 32-M-Rh-N3** anchored at pH = 5 (**FM5**) and at pH = 3 (**FM3**) (Figure 7). As previously commented (Figure 2), the micro- and nanoparticles exhibited differences with regard to size and shape (Figure 5a,b). The micro MSPs were irregular in shape and size, whereas the nano MSPs appeared as equal spheres of ca. 100 nm. These particle differences conditioned the way in which the MSPs anchored on the films. Figure 5c,d shows the distribution of micro- and nanoparticles on the EVOH 32 film surface UV-treated for 15 min after reaction of the EVOH 32 film with an equal mass of **M-Rh-N3** or **N-Rh-N3**. A much better dispersion was observed for the nanoparticles (Figure 5d) when compared with the microparticles (Figure 5c). Moreover, the number of particles attached to the surface of the film was estimated as 0.05 ± 0.01 particles/μ^2 for **M-Rh-N3** and 67 ± 5 particles/μ^2 for **N-Rh-N3**. Similar results, although with a much lesser immobilization of particles, were obtained for the EVOH 44 films (data not shown). Therefore, the EVOH 44 films were discarded.

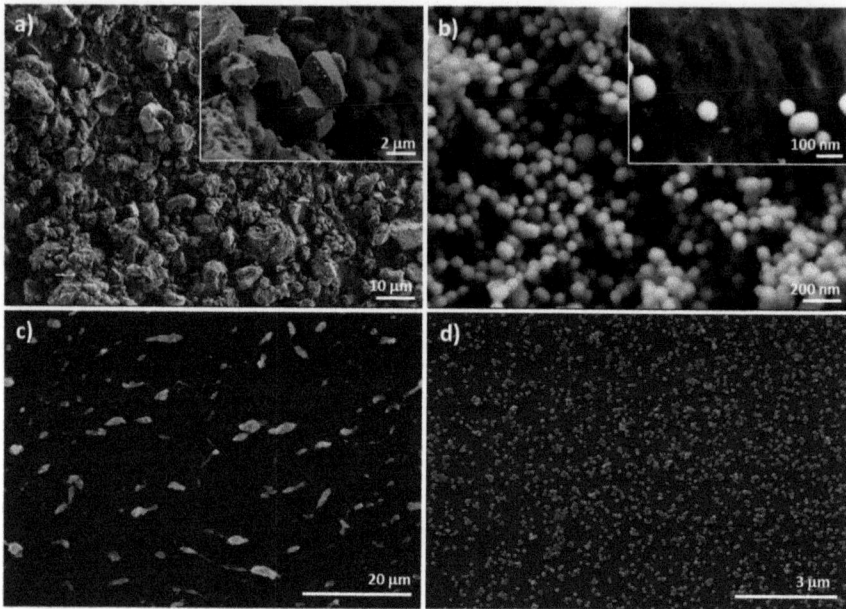

Figure 5. Representative field-emission SEM (FESEM) images of (**a**) **M-Rh-N3** and (**b**) **N-Rh-N3** particles, and (**c**) **M-Rh-N3** and (**d**) **N-Rh-N3** particles covalently anchored to EVOH 32 film after being UV-treated for 15 min.

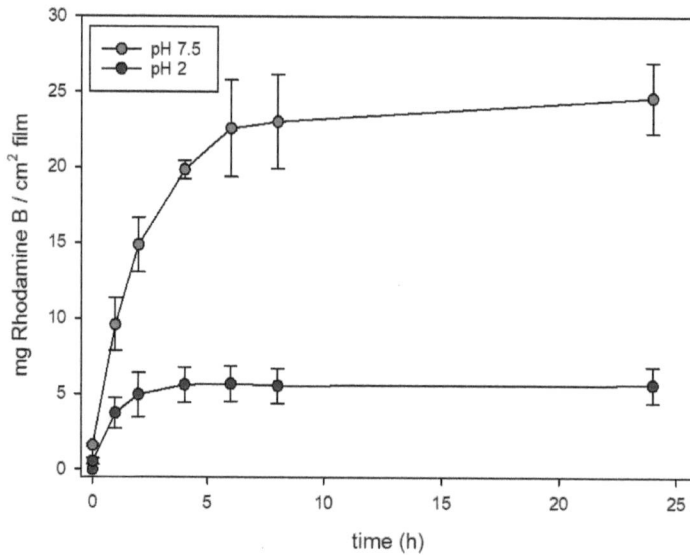

Figure 6. Evolution of rhodamine B release from **EVOH 32-N-Rh-N3** prepared at pH 3 in aqueous media at acidic pH (pH = 2) and at neutral pH (pH = 7.5).

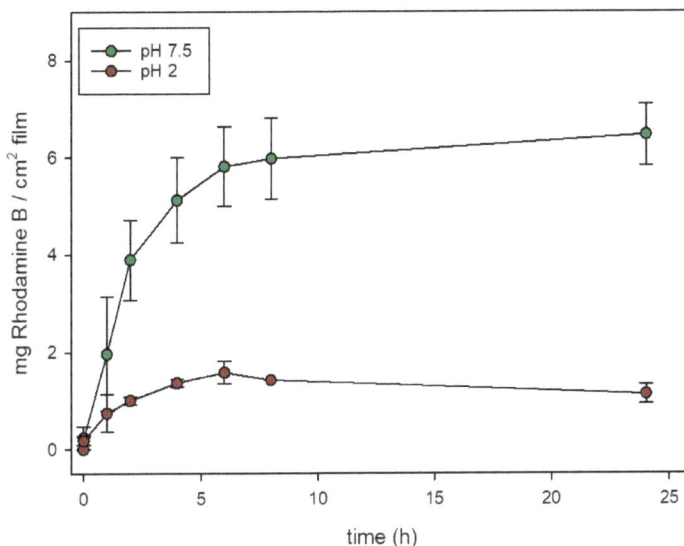

Figure 7. Evolution of rhodamine B release from **EVOH 32-M-Rh-N3** prepared at pH 3 in aqueous media at acidic pH (pH = 2) and at neutral pH (pH = 7.5).

3.4. Controlled Release Behavior

Finally, and with the objective of confirming the efficiency of the gated materials for releasing rhodamine B, controlled delivery studies from the functionalized films according to the pH in water (used as an aqueous food simulant) were carried out.

Figures 6 and 7 show the release rates of rhodamine B from **EVOH 32-N-Rh-N3 and EVOH 32-M-Rh-N3** films functionalized at pH 3 (expressed as mg of rhodamine B released from 1 cm^2 of film) when immersed in two aqueous food simulants, at acidic (pH = 2) and neutral pH (pH = 7.5). In **EVOH 32-N-Rh-N3** films (Figure 6), the maximum release of rhodamine B was observed at neutral pH. In these conditions, a controlled, sustained release was achieved during 8 h reaching a maximum delivery content of ca. 24.5 mg of rhodamine B per cm^2 of film Conversely, when this sample was exposed to an acidic medium, the release was significantly reduced (maximum delivery of ca. 5.5 mg of rhodamine B per cm of film), reaching a flat baseline during the first hours of the experiment. This different and remarkable behavior at pH 7.5, when compared to that at pH 2, was due to the effect of pH on the conformation of the polyamines. This gating-mechanism was widely described in recent years [33]. At pH 2, polyamines are protonated into polyammonium groups that favor Coulombic repulsions between close chains. Tethered polyammonium moieties tend to adopt a rigid-like conformation that pushes them away toward pore openings, blocking the pores and completely or partially inhibiting release of the sorbed substance. In contrast, a progressive delivery of the colorant was observed at pH 7.5. In this condition (neutral pH), polyamines are less protonated and their repulsions are weaker, favoring a pore unblockage that allows the release of the encapsulated fluorophore.

Similar behavior can be seen in Figure 7 for **EVOH 32-M-Rh-N3**. A sustained release of the cargo was observed after adding water adjusted to pH 7.5 to the films, while release was hindered after the addition of water adjusted to pH 2. Again, at neutral pH, polyamines are not protonated and their interactions are weaker, favoring pore unlock. However, despite the fact that MCM-41 nano- and microparticles were loaded with similar amounts of rhodamine B (see Table 1), the amount of rhodamine delivered from **EVOH 32-M-Rh-N3** was ca. four-fold lower than that achieved from

EVOH 32-N-Rh-N3 films. This difference might be explained by the different density of particles anchored to the film (see Figure 4).

Finally, the results presented in this work also confirm that amines do not lose their gating properties after immobilization on EVOH films, opening a new field for controlled release in packaging applications.

To compare the rhodamine B release kinetics from pore voids of both types of silica supports (nano and micro) the experimental release data were fitted to the Higuchi model, and the Higuchi release rate constant (k_H) was calculated. The good fit of the delivery curves to the Higuchi model, as Figure 8 shows, suggests that the delivery of rhodamine B from the pores of various solids is basically a diffusive process [31]. Moreover, for both types of particles, the k_H constant at pH 7.5 was the same ($k_H = 40$). There were also no significant differences in Higuchi rate constants at acidic conditions: k_H was 15 for **EVOH 32-N-Rh-N3**, and 13 for **EVOH 32-M-Rh-N3**. These data confirm that delivery at pH 7.5 is not only more efficient than at pH 2, but also faster. Moreover, the small differences between the two types of particles demonstrate that release kinetics is influenced by the porous system (equal in all MCM-41 particles) instead of by the particle morphology. Accordingly, differences observed in the comparison of Figure 6 with regard to the amount of dye released by the encapsulation systems (20%–30% greater in **EVOH 32-N-Rh-N3** films) are due to the higher concentration of particles achieved during the N3-MSP deposition step (see Figure 4).

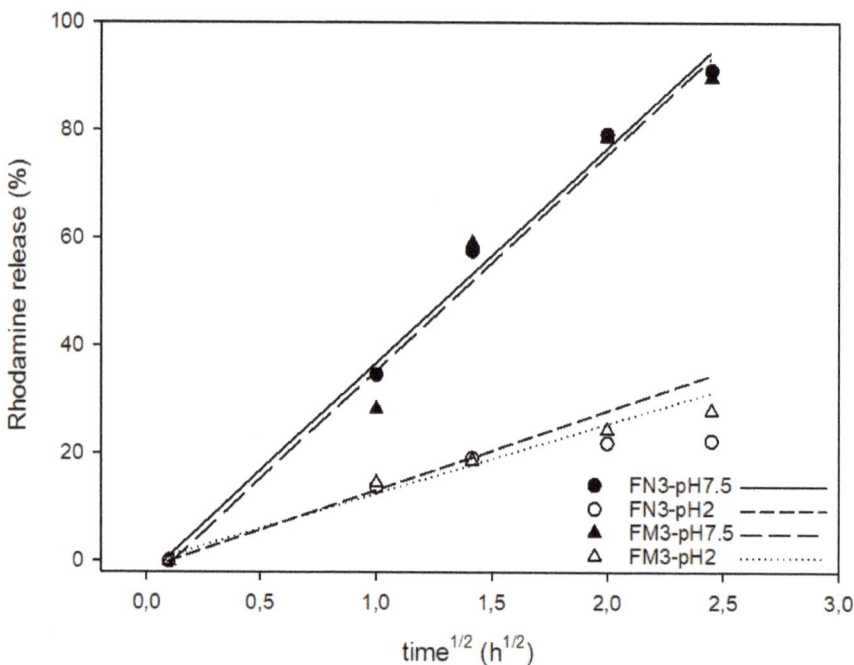

Figure 8. Higuchi representation of rhodamine B release from **EVOH 32-M-Rh-N3** and **EVOH 32-N-Rh-N3** prepared at pH 3 in aqueous media at acidic pH (pH = 2) and at neutral pH (pH = 7.5). Symbols correspond to experimental data, and the lines are the fitting lines of the Higuchi equation.

4. Conclusions

This work is a proof of concept for the design of active packaging materials with a pH triggering mechanism. Mesoporous silica micro- and nanoparticles (MCM-41) were manufactured, calcined, and used to load and release an agent in a controlled manner when exposed to suitable pH conditions. The particles were loaded with rhodamine B (selected as a dye whose release is easy to monitor) and functionalized with *N*-(3-trimethoxysilylpropyl)diethylenetriamine (polyamines). This functionalization creates chemical gates whose key is based on pH.

In parallel, poly(ethylene-*co*-vinyl alcohol) (EVOH) films with two monomer compositions, EVOH 32 and EVOH 44, were successfully oxidized by UV irradiation. The treatment generated –COOH substituents in the polymer chains, which increased with an increase in irradiation time and a decrease in copolymer ethylene content. Oxidized EVOH was used to anchor the loaded silica particles through the use of EDC/NHS linkers. Linkage was carried out successfully at pH 3 and pH 5, especially for nanoparticles, which were distributed homogeneously throughout the film surface, especially in the case of EVOH 32 films.

Finally, the ability to keep and release the agent was analyzed. The final load of the dye was greater in the films exposed to anchorage treatments at pH 3, as, at pH 5, a partial release of rhodamine B was evidenced during the process. The films with the anchored particles were exposed to two liquid media, simulating acidic food and neutral food. The films released the agent quickly and completely at neutral pH, but kept the dye locked at acidic pH.

Hence, this work demonstrates the feasibility of covalently anchoring smart delivery systems able to deliver a functional molecule after applying a triggering stimulus to EVOH films. This successful mechanism will allow the design of new active packaging systems loaded with antioxidant, antimicrobial, or aromatic agents able to release their cargo only under certain conditions, such as the generation of biogenic amines by bacteria in fresh food products.

Author Contributions: Investigation, V.M.-G., É.P.-E., M.R.-R., R.M.-M., J.M.B., P.H.-M., and R.G.

Funding: This research was funded by the Ministry of Economy, Industry, and Competitiveness, projects AGL2015-64595-R, AGL2015-70235-C2-1-R, and AGL2015-70235-C2-2-R.

Acknowledgments: The authors acknowledge the assistance of Karel Clapshaw (translation services).

Conflicts of Interest: The authors declare no conflicts of interest.

References

1. Appendini, P.; Hotchkiss, J.H. Review of antimicrobial food packaging. *Innov. Food Sci. Emerg. Technol.* **2002**, *3*, 113–126. [CrossRef]
2. Larson, A.M.; Klibanov, A.M. Biocidal packaging for pharmaceuticals, foods, and other perishables. *Annu. Rev. Chem. Biomol. Eng.* **2013**, *4*, 171–186. [CrossRef] [PubMed]
3. Yam, K.L.; Lee, D.S. *Emerging Food Packaging Technologies: Principles and Practice*; Woodhead Publishing: Sawston, UK, 2012.
4. Gavara, R.; López-Carballo, G.; Hernández-Muñoz, P.; Catalá, R.; Muriel-Galet, V.; Cerisuelo, J.P.; Dominguez, I. *Practical Guide to Antimicrobial Active Packaging*; Smithers Rapra Publishing: Surrey, UK, 2015; p. 263.
5. Otoni, C.G.; Espitia, P.J.P.; Avena-Bustillos, R.J.; McHugh, T.H. Trends in antimicrobial food packaging systems: Emitting sachets and absorbent pads. *Food Res. Int.* **2016**, *83*, 60–73. [CrossRef]
6. Lopez-Rubio, A.; Almenar, E.; Hernandez-Munoz, P.; Lagaron, J.M.; Catala, R.; Gavara, R. Overview of active polymer-based packaging technologies for food applications. *Food Rev. Int.* **2004**, *20*, 357–387. [CrossRef]
7. Charles, F.; Sanchez, J.; Gontard, N. Absorption kinetics of oxygen and carbon dioxide scavengers as part of active modified atmosphere packaging. *J. Food Eng.* **2006**, *72*, 1–7. [CrossRef]
8. Cerisuelo, J.P.; Muriel-Galet, V.; Bermudez, J.M.; Aucejo, S.; Catala, R.; Gavara, R.; Hernandez-Munoz, P. Mathematical model to describe the release of an antimicrobial agent from an active package constituted by carvacrol in a hydrophilic evoh coating on a pp film. *J. Food Eng.* **2012**, *110*, 26–37. [CrossRef]
9. Chalier, P.; Ben Arfa, A.; Guillard, V.; Gontard, N. Moisture and temperature triggered release of a volatile active agent from soy protein coated paper: Effect of glass transition phenomena on carvacrol diffusion coefficient. *J. Agric. Food Chem.* **2009**, *57*, 658–665. [CrossRef] [PubMed]
10. Bushman, A.C.; Castle, G.J.; Pastor, R.D.; Reighard, T.S. Process for Activating Oxygen Scavenger Components during a Gable-Top Carton Filling Process. U.S. Patent 6,689,314B2, 10 February 2004.

11. Goddard, J.M.; Talbert, J.N.; Hotchkiss, J.H. Covalent attachment of lactase to low-density polyethylene films. *J. Food Sci.* **2007**, *72*, E36–E41. [CrossRef] [PubMed]

12. Muriel Galet, V.; Talbert, J.N.; Hernandez Munoz, P.; Gavara, R.; Goddard, J.M. Covalent immobilization of lysozyme on ethylene vinyl alcohol films for nonmigrating antimicrobial packaging applications. *J. Agric. Food Chem.* **2013**, *61*, 6720–6727. [CrossRef] [PubMed]

13. Saini, S.; Sillard, C.; Belgacem, M.N.; Bras, J. Nisin anchored cellulose nanofibers for long term antimicrobial active food packaging. *RSC Adv.* **2016**, *6*, 12437–12445. [CrossRef]

14. Roman, M.J.; Decker, E.A.; Goddard, J.M. Biomimetic polyphenol coatings for antioxidant active packaging applications. *Colloid Interface Sci. Commun.* **2016**, *13*, 10–13. [CrossRef]

15. Vasile, C.; Pâslaru, E.; Sdrobis, A.; Pricope, G.; Ioanid, G.E.; Darie, R.N. Plasma assisted functionalization of synthetic and natural polymers to obtain new bioactive food packaging materials. In Proceedings of the Application of Radiation Technology in Development of Advanced Packaging Materials for Food Products Vienna, Austria, 22–26 April 2013; Safrani, A., Ed.; pp. 100–110.

16. Mihindukulasuriya, S.D.F.; Lim, L.T. Nanotechnology development in food packaging: A review. *Trends Food Sci. Technol.* **2014**, *40*, 149–167. [CrossRef]

17. Cerisuelo, J.P.; Alonso, J.; Aucejo, S.; Gavara, R.; Hernandez-Munoz, P. Modifications induced by the addition of a nanoclay in the functional and active properties of an evoh film containing carvacrol for food packaging. *J. Membr. Sci.* **2012**, *423–424*, 247–256. [CrossRef]

18. Biddeci, G.; Cavallaro, G.; Di Blasi, F.; Lazzara, G.; Massaro, M.; Milioto, S.; Parisi, F.; Riela, S.; Spinelli, G. Halloysite nanotubes loaded with peppermint essential oil as filler for functional biopolymer film. *Carbohydr. Polym.* **2016**, *152*, 548–557. [CrossRef] [PubMed]

19. Li, W.; Zhang, C.; Chi, H.; Li, L.; Lan, T.; Han, P.; Chen, H.; Qin, Y. Development of antimicrobial packaging film made from poly(lactic acid) incorporating titanium dioxide and silver nanoparticles. *Molecules* **2017**, *22*, 1170. [CrossRef] [PubMed]

20. Mal, N.K.; Fujiwara, M.; Tanaka, Y.; Taguchi, T.; Matsukata, M. Photo-switched storage and release of guest molecules in the pore void of coumarin-modified mcm-41. *Chem. Mater.* **2003**, *15*, 3385–3394. [CrossRef]

21. Mal, N.K.; Fujiwara, M.; Tanaka, Y. Photocontrolled reversible release of guest molecules from coumarin-modified mesoporous silica. *Nature* **2003**, *421*, 350–353. [CrossRef] [PubMed]

22. Hernandez, R.; Tseng, H.R.; Wong, J.W.; Stoddart, J.F.; Zink, J.I. An operational supramolecular nanovalve. *J. Am. Chem. Soc.* **2004**, *126*, 3370–3371. [CrossRef] [PubMed]

23. Casasus, R.; Marcos, M.D.; Martinez-Manez, R.; Ros-Lis, J.V.; Soto, J.; Villaescusa, L.A.; Amoros, P.; Beltran, D.; Guillem, C.; Latorre, J. Toward the development of ionically controlled nanoscopic molecular gates. *J. Am. Chem. Soc.* **2004**, *126*, 8612–8613. [CrossRef] [PubMed]

24. Aznar, E.; Martínez-Máñez, R.; Sancenón, F. Controlled release using mesoporous materials containing gate-like scaffoldings. *Expert Opin. Drug Deliv.* **2009**, *6*, 643–655. [CrossRef] [PubMed]

25. Hung, B.-Y.; Kuthati, Y.; Kankala, R.K.; Kankala, S.; Deng, J.-P.; Liu, C.-L.; Lee, C.-H. Utilization of enzyme-immobilized mesoporous silica nanocontainers (IBN-4) in prodrug-activated cancer theranostics. *Nanomaterials* **2015**, *5*, 2169–2191. [CrossRef] [PubMed]

26. Kankala, R.K.; Liu, C.-G.; Chen, A.-Z.; Wang, S.-B.; Xu, P.-Y.; Mende, L.K.; Liu, C.-L.; Lee, C.-H.; Hu, Y.-F. Overcoming multidrug resistance through the synergistic effects of hierarchical pH-sensitive, ros-generating nanoreactors. *ACS Biomater. Sci. Eng.* **2017**, *3*, 2431–2442. [CrossRef]

27. Huang, P.-K.; Lin, S.-X.; Tsai, M.-J.; Leong, M.K.; Lin, S.-R.; Kankala, R.K.; Lee, C.-H.; Weng, C.-F. Encapsulation of 16-hydroxycleroda-3,13-dine-16,15-olide in mesoporous silica nanoparticles as a natural dipeptidyl peptidase-4 inhibitor potentiated hypoglycemia in diabetic mice. *Nanomaterials* **2017**, *7*, 112. [CrossRef] [PubMed]

28. Kuthati, Y.; Kankala, R.K.; Busa, P.; Lin, S.-X.; Deng, J.-P.; Mou, C.-Y.; Lee, C.-H. Phototherapeutic spectrum expansion through synergistic effect of mesoporous silica trio-nanohybrids against antibiotic-resistant gram-negative bacterium. *J. Photochem. Photobiol. B Biol.* **2017**, *169*, 124–133. [CrossRef] [PubMed]

29. Sancenon, F.; Pascual, L.; Oroval, M.; Aznar, E.; Martinez-Manez, R. Gated silica mesoporous materials in sensing applications. *Chemistryopen* **2015**, *4*, 418–437. [CrossRef] [PubMed]

30. Kankala, R.K.; Kuthati, Y.; Liu, C.-L.; Mou, C.-Y.; Lee, C.-H. Killing cancer cells by delivering a nanoreactor for inhibition of catalase and catalytically enhancing intracellular levels of ros. *RSC Adv.* **2015**, *5*, 86072–86081. [CrossRef]

31. Perez-Esteve, E.; Ruiz-Rico, M.; de la Torre, C.; Villaescusa, L.A.; Sancenon, F.; Marcos, M.D.; Amoros, P.; Martinez-Manez, R.; Manuel Barat, J. Encapsulation of folic acid in different silica porous supports: A comparative study. *Food Chem.* **2016**, *196*, 66–75. [CrossRef] [PubMed]

32. Kumar, S.; Raju, S.; Mohana, N.; Sampath, P.; Jayakumari, L. Effects of nanomaterials on polymer composites—An expatiate view. *Rev. Adv. Mater. Sci.* **2014**, *38*, 40–54.

33. Perez-Esteve, E.; Fuentes, A.; Coll, C.; Acosta, C.; Bernardos, A.; Amoros, P.; Marcos, M.D.; Sancenon, F.; Martinez-Manez, R.; Barat, J.M. Modulation of folic acid bioaccessibility by encapsulation in pH-responsive gated mesoporous silica particles. *Microporous Mesoporous Mater.* **2015**, *202*, 124–132. [CrossRef]

34. Uchida, E.; Uyama, Y.; Ikada, Y. Sorption of low-molecular-weight anions into thin polycation layers grafted onto a film. *Langmuir* **1993**, *9*, 1121–1124. [CrossRef]

35. Kang, E.T.; Tan, K.L.; Kato, K.; Uyama, Y.; Ikada, Y. Surface modification and functionalization of polytetrafluoroethylene films. *Macromolecules* **1996**, *29*, 6872–6879. [CrossRef]

36. Perez-Esteve, E.; Ruiz-Rico, M.; Martinez-Manez, R.; Manuel Barat, J. Mesoporous silica-based supports for the controlled and targeted release of bioactive molecules in the gastrointestinal tract. *J. Food Sci.* **2015**, *80*, E2504–E2516. [CrossRef] [PubMed]

37. Zhou, H.; Wang, X.; Tang, J.; Yang, Y.-W. Surface immobilization of pH-responsive polymer brushes on mesoporous silica nanoparticles by enzyme mimetic catalytic atrp for controlled cargo release. *Polymers* **2016**, *8*, 277. [CrossRef]

38. Lazzara, G.; Cavallaro, G.; Panchal, A.; Fakhrullin, R.; Stavitskaya, A.; Vinokurov, V.; Lvov, Y. An assembly of organic-inorganic composites using halloysite clay nanotubes. *Curr. Opin. Colloid Interface Sci.* **2018**, *35*, 42–50. [CrossRef]

39. Lisuzzo, L.; Cavallaro, G.; Lazzara, G.; Milioto, S.; Parisi, F.; Stetsyshyn, Y. Stability of halloysite, imogolite, and boron nitride nanotubes in solvent media. *Appl. Sci.* **2018**, *8*, 1068. [CrossRef]

40. Bryce-Smith, D.; Gilbert, A. Photodegradation of polymers. In *Photochemistry*; Royal Society of Chemistry: Cambridge, UK, 1993; pp. 448–458.

© 2018 by the authors. Licensee MDPI, Basel, Switzerland. This article is an open access article distributed under the terms and conditions of the Creative Commons Attribution (CC BY) license (http://creativecommons.org/licenses/by/4.0/).

nanomaterials

MDPI

Article

Investigation into the Potential Migration of Nanoparticles from Laponite-Polymer Nanocomposites

Johannes Bott * and Roland Franz

Department of Product Safety and Analytics, Fraunhofer Institute for Process Engineering and Packaging (IVV), 85354 Freising, Germany; roland.franz@ivv.fraunhofer.de
* Correspondence: johannes.bott@ivv.fraunhofer.de; Tel.: +49-8161-491-753

Received: 13 August 2018; Accepted: 10 September 2018; Published: 13 September 2018

Abstract: In this study, the migration potential of laponite, a small synthetic nanoclay, from nanocomposites into foods was investigated. First, a laponite/ethylene vinyl acetate (EVA) masterbatch was compounded several times and then extruded into thin low-density polyethylene (LDPE) based films. This way, intercalation and partial exfoliation of the smallest type of clay was achieved. Migration of laponite was investigated using Asymmetric Flow Field-Flow Fractionation (AF4) with Multi-Angle Laser Light Scattering (MALLS) detection. A surfactant solution in which laponite dispersion remained stable during migration test conditions was used as alternative food simulant. Sample films with different loadings of laponite were stored for 10 days at 60 °C. No migration of laponite was found at a limit of detection of 22 µg laponite per Kg food. It can be concluded that laponite (representing the worst case for any larger structured type of clay) does not migrate into food once it is incorporated into a polymer matrix.

Keywords: laponite; clay; nanomaterial; migration; diffusion; nanocomposites

---

1. Introduction

Laponite is a synthetic colloidal layered silicate which is composed of disc-shaped crystals in the nanoscale size region. Laponite forms the same tetrahedral and octahedral 2:1 "sandwich" structure like most natural clays, e.g., montmorillonite [1]. In contrast, the primary structure (i.e., the disc-shaped crystals) is significant smaller than it is in naturally occurring clay minerals. Laponite crystals are of approximately 1 nm in thickness and have diameters of approximately 25 nm, only [2]. In general, clays are used as polymer additives nearly as long as polymers are on the market. From the beginning, clays are used as matrix fillers improving thermal and mechanical properties of the polymer significantly. In its exfoliated form, clay promises to enhance the barrier function plastic food contact materials (FCM) regarding oxygen or water vapor [3–20]. Laponites are mainly used as a rheology modifier or as film formers. They find applications in many consumer care products to improve suspension stability (e.g., cleaning products) and to adjust rheological behaviour (e.g., used as thickener in toothpaste). Laponites are also applied as a thin film on the surface of different materials like paper and polymeric films to produce food packaging that are antistatic or that have improved barrier properties regarding oxygen transmission [21]. Furthermore, the successful use of laponite disks in sustainable polymers like hydroxypropyl cellulose [22] and pectins [23] was demonstrated.

Due to its structural composition of ultrafine platelets with only 1 nm in thickness, laponite (and clay) has to be considered as a nanomaterial (NM) according to the definition of the European Union 2011/696/EU [24]. Nanomaterials in general promise many technical and economic benefits and are increasingly used. Therefore, an assessment of these materials from a safety perspective is always needed. This is, in particular, the case when such NMs are used as additives in FCM. It is generally accepted that a risk for the consumer is only given when exposure to NMs used as a polymer additive for food packaging is performed. In case of nanocomposites, exposure towards the NM would only be the case if the NM migrates out of the packaging into food. Within the last years the number of studies that investigated the migration potential of different types of NMs has increased and can be found summarized elsewhere [7,25,26]. In some case-examples practical evidence was already given that some NMs cannot migrate and theoretical considerations even showed that NMs used as additives in FCM are too large to migrate in general [26,27]. The use of NMs as additives in FCM made of plastics is regulated within the European Union in regulation EU/10/2011 [28] and its amendments [29–31]. Legal requirements demand that a risk assessment of NMs in FCM has to be performed on a case-by-case basis.

In this study, the migration potential of laponite from nanocomposites based on low-density polyethylene (LDPE) was investigated. Due to its small size compared to other nanoclays, laponite was selected as a model nanomaterial which will represent the worst-case in comparison with all other types of clay materials when considering migration being based on size-dependent Fickian diffusion [27].

2. Materials and Methods

2.1. Materials

Organically modified laponite RXG7308 (in the following: Laponite) (BYK Chemie GmbH, Moosburg, Germany) was provided both as a pure powder and also incorporated into ethylene-vinyl acetate (EVA) as a masterbatch (SO 9015, BYK Chemie GmbH, Moosburg, Germany) with a laponite content of 11%-m/m. The powder was used for analytical method development and the masterbatch was used to produce LDPE films with different contents of laponite for migration measurements.

The production of the LDPE films was performed in-house at Fraunhofer IVV. For this the SO 9015 masterbatch was first mixed with neat ethylene-vinyl acetate copolymer (Escorene Ultra FL 00226CC, 26% vinyl acetate, ExxonMobil Chemical Company, Houston, TX, USA) and compounded six times using a twin-screw compounder (Dr. Collin GmbH). With the preceded compounding strong shear forces are applied to the laponite stacks which shall provide better homogeneity in the LDPE films and partial intercalation/exfoliation of the layered laponite stacks. After compounding, the EVA/laponite masterbatch was mixed with LDPE (Lupolen 1806 H, LyondellBasell, Rotterdam, The Netherlands) and extruded to films with 2%, 4% and 6% laponite in the polymer using a flat film extruder (Dr. Collin GmbH). An LDPE film without laponite in the polymer was extruded as reference film.

2.2. Transmission Electron Microscopy (TEM)

TEM images of the polymeric film with the lowest laponite concentration were prepared by psi cube, Germany. With this technique the distribution and size characteristics of the laponite in the polymer was visualized. For sample preparation the polymeric film was subjected to cryo-ultra-thin-sectioning using a diamond knife.

2.3. Preparation of Laponite Reference Dispersions

Ultrapure water (TKA Genpure, Fisher Scientific GmbH, Schwerte, Germany) with 25000 mg/L of the surfactant Novachem (Postnova Analytics) and 200 mg/L of the biocide sodium azide (Merck Millipore, Darmstadt, Germany) was used as dispersant for a laponite stock dispersion. Dilutions of the initial stock dispersion were produced using ultrapure water with 2000 mg/L of the surfactant only. Ultrapure water with 200 mg/L sodium azide, without surfactants, was used as flowing liquid for the AF4. After preparation all solvents were filtered (0.1 μm Millipore filter disc).

For the stock dispersion 50.0 mg Laponite RXG7308 powder (BYK Chemie GmbH, Moosburg, Germany) was weighed out into a 50 mL polypropylene centrifuge vial and mixed with 20 mL of the dispersant. The centrifuge vial was then placed into an ice bath and the mixture was dispersed for 45 min using an ultra-sonication tip (Vibra Cell VC 50T, Sonics&Materials Inc., Newton, CT, USA, operated at 50 Watt, 20 kHz, 100% output), which was placed approximately 1 cm above the bottom of the vial. At the end of the dispersion experiment the dispersion was transferred quantitatively into a 200 mL perfluoroalkoxy alkane (PFA) measuring flask and filled up to the mark. This way a laponite dispersion with a nominal laponite concentration of 250 mg/L was prepared. The dispersion did neither showed sedimentation of laponite nor a strong milky turbidity, but was slightly opalescent. A complete exfoliation of the Laponite layers would result in clear dispersions, wherefore the slight opalescence was an indication that Laponite aggregates were broken into smaller units and at least intercalation of the Laponite stacks was successful, due to the high energy input of the ultra-sonication tip.

2.4. AF4 and MALLS Measurements

AF4 measurements were carried out with an "AF2000 MT Series mid temperature" (Postnova Analytics, Landsberg, Germany) to characterize, detect and quantify laponite particles in the ongoing migration experiments. The system was equipped with a 500 µm channel and a polyethersulfone membrane (cut-off: 10 kDa). For the determination of the particles size distribution a 21-angle-MALLS detector "PN3621" (Postnova Analytics, Landsberg, Germany) was used.

For diluted Laponite RXG7308 dispersions the AF4 conditions were optimised as follows: During the injection time of 15 min (this is tantamount to the focusing time), the cross flow is kept constant at 0.8 mL/min (start conditions). Within a transition time of 0.5 min the focusing of the sample is terminated and the elution of sample starts. The cross flow is kept constant for additional 5 min followed by a fast non-linear decline of the cross flow to 0 mL/min within 10.0 min, using a power gradient of 0.15. The channel is flushed by the detector flow for 35 min without any cross flow to deplete the channel completely. The detector flow is kept constant at 0.45 mL/min for the whole run. The channel was tempered to constant 40 °C. The method was used for sample injection volumes up to 2000 µL.

For laponite particles, a serial dilution from a 250 mg/L stock solution was made and standard solutions with 0 ng/mL (blank), 250 ng/mL, 500 ng/mL, 1000 ng/mL, 1500 ng/mL, 2000 ng/mL and 2500 ng/mL were measured. The detected signals of each standard dispersion were integrated at all detection angles (excluding the 7°, 12° and 164° angle) of the MALLS detector to obtain the peak area of the sample. The detected peak area is directly proportional to the injected mass.

Furthermore, the results of the light scattering experiment were used to calculate the sizes of dispersed laponite particles in form of the radius of gyration, r_g.

2.5. Migration Test

For the migration studies simulants and test conditions were intended to be chosen according to Annex V of the European Plastics Regulation (EU) 10/2011 for long term storage at room temperature (more than 6 month) including hotfill (2 h at 70 °C or 15 min at 100 °C). Preliminary dispersion experiments showed that only in the 2000 mg/L Novachem surfactant solution sufficient dispersion stability of laponite particles could be achieved. Therefore, the surfactant solution (2000 mg/L Novachem and 200 mg/L sodium azide) was chosen as (alternative) food simulant.

Sample films were cut into squares to a defined area of 1 dm^2. The samples were stored in 30 mL polypropylene vials. Both samples and vials were blown out before with nitrogen to prevent contamination by dust, filled with 15 mL of the food simulant and stored for 10 days at 60 °C. All samples were completely covered with the simulant during storage. From each sample film containing laponite in the polymer (2%, 4%, 6% Laponite in LDPE) four equally treated samples were prepared, from the LDPE blanks three identical samples. All migration samples were injected into the AF4/MALLS system twice

Additionally, LDPE reference samples (without Laponite) and solvent blanks were prepared and stored under the same conditions.

To validate the migration experiments tests on recovery were performed. Freshly prepared laponite dispersions with 1000 ng/mL laponite were stored for 10 days at 60 °C in 2000 mg/L Novachem solution in the same vials as described above. Stored dispersions were measured by AF4/MALLS and compared to a measurement performed directly after preparation of the Laponite dispersion in a Novachem solution. The recovery rates are calculated as the ratio of the detected total peak areas (MALLS output of all detector angles except the 7° angle).

3. Results

3.1. TEM Measurements

The lower resolution (Figure 1a) gives a picture of the laponite distribution in the polymer showing that the laponite exists as layered stacks which are homogenously distributed in the polymer. However, a statistical evaluation of the particle size distribution of laponite within the polymer was not performed by TEM. The high resolution image (Figure 1b) shows single laponite aggregates in the polymer. At this resolution, it can be seen that the platelets of laponite do not form compact stacks anymore but are rather oriented randomLy and not parallel to each other. At the border region of the laponite aggregates single platelets were found. This was an indication that the Laponite stacks were already intercalated and partial exfoliation of the stacks took place.

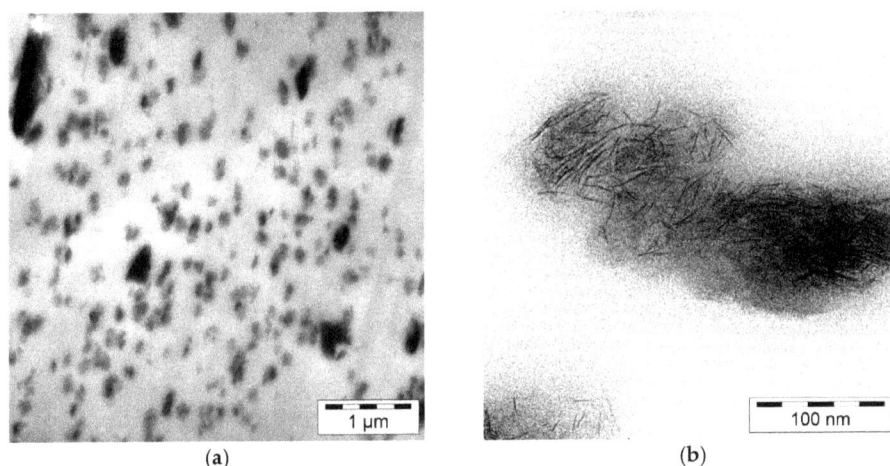

(a) (b)

Figure 1. Transmission Electron Microscopy (TEM) images of the 6% laponite in LDPE film: (a) low and (b) high resolved image.

3.2. Characterization and Quantification of Laponite in Dispersion

Fractograms of laponite reference dispersions and the solvent blank are shown in Figure 2. Beside the signal caused by the Laponite particles, the 2000 mg/L Novachem dispersant blank caused a signal at elution times typical for the Laponite particles ($t = 21$ to 40 min). This is caused by a pressure drop within the AF4 channel during the fast decline of the cross flow. At an injection volume of 2000 μL each sample at a concentration between 250 ng/mL and 2500 ng/mL delivered a signal that could be distinguished from the blank and from samples with other concentrations. A 250 ng/mL laponite sample still delivered an evaluable signal. Concentrations lower than 250 ng/mL lead to problems for an explicit evaluation by light scattering. Thus, the lowest detectable concentration for laponite particles with this AF4 system and method is 250 ng/mL. At the injection volume of 2000 μL this corresponds to 500 ng of laponite.

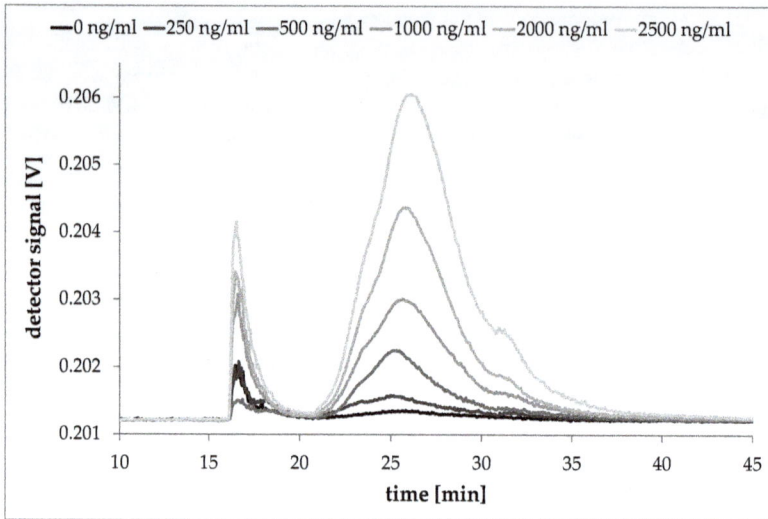

Figure 2. Serial dilution of laponite dispersion in 2000 mg/L Novachem solution (2000 µL injected, signal of the 92° detector).

The signal outputs (excluding 7°, 12° and 164° detector angle) of the MALLS detector from the respective laponite reference dispersions were integrated, summed up to the total peak area and correlated with the concentration of the respective laponite reference dispersion (Figure 3). The function of this correlation experiment was used to determine laponite concentrations in unknown samples. The results of the calibration experiment are summarized in Table 1.

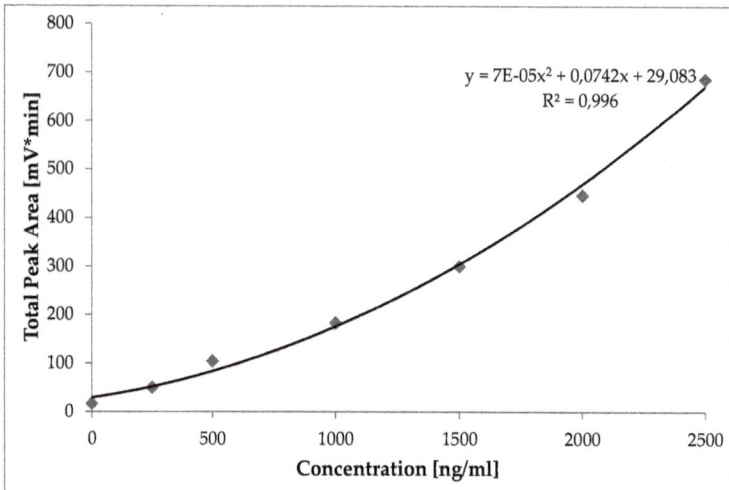

Figure 3. Sum of all MALLS detector angle areas (total area) of the laponite peaks versus the concentration (injection volume 2000 µL).

Table 1. Peak areas of Laponite particles obtained by Multi Angle Laser Light Scattering (MALLS) detection (2000 µL injection volume).

Concentration of Standard (ng/mL)	Mass (ng)	Total Area by MALLS (mV*min)
0	0	16.7
250	500	50.2
500	1000	105.0
1000	2000	183.7
1500	3000	300.1
2000	4000	446.9
2500	5000	686.3

Via the MALLS detection, the particle sizes of laponite particles were determined in dispersion. For the calculation of the radius of gyration a random coil fit was used. In Figure 4 the fractogram (signal intensity vs. elution time) of a 2 mg/L laponite dispersion is overlaid by the calculated radii at the respective elution times. The almost linear particle radius increase with increasing elution volume indicates a successful separation of the laponite particles. The laponite particles show a particle size distribution starting from about 16 nm to about 130 nm (radius of gyration r_g). For the main part of the particles the calculated radius of gyration was about 41 nm. Radius of gyration, also root mean square radius, describes the distribution of mass around the centre of mass of the particular particle. This calculation is based on the angular variation of the signal intensities of the MALLS detector. The calculation of the radius of gyration is independent of the shape of the particle. However, r_g can be re-calculated into geometrical sizes under the assumption of specific geometrical shapes [32,33]. In case of random coils, the geometrical end-to-end distance the particle size distribution ranges from 21 nm to 168 nm with the main part with 53 nm. This indicates that both, small primary units as well as smaller stacks of laponite were present in dispersion. This way, the dispersion used for AF4/MALLS method development covered the same laponite sizes as it was found in TEM images of the nanocomposites which were used in the later migration experiment.

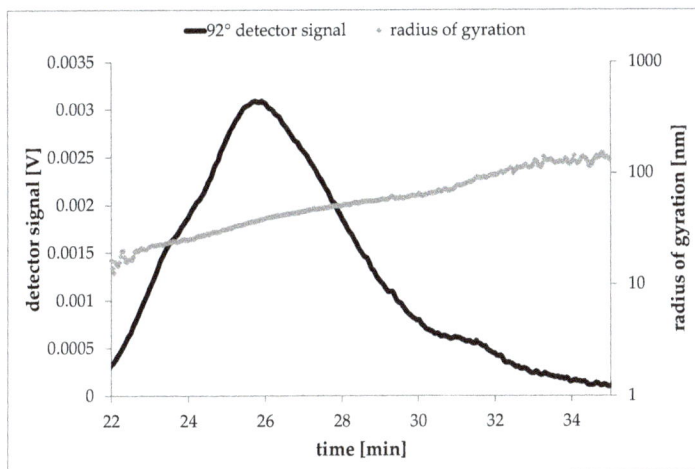

Figure 4. Elugram of an AF4 run with laponite particles (2000 ng/mL, 2000 µL injected. Signal of the 92° detector overlaid with the calculated radii of gyration at the relevant elution times.

3.3. Migration Test Results

All migration experiments were carried out using a 2000-mg/L Novachem solution as simulant. Laponite particles are expected to elute from about $t = 21$ to 40 min. Due to the rapid decrease of the cross flow at that time the surfactant blank caused a slight signal (see Figure 2). AF4/MALLS measurements of the migration sample without laponite in the polymer (Figure 5a) and with 2% (Figure 5b), 4% (Figure 5c) and 6% laponite (Figure 5d) in the polymer showed, that no higher signal intensities than the surfactant blank could be detected in any sample, indicating that no oligomers were extracted and no laponite particles migrated into the food simulant after storage for 10 days at 60 °C. Rather, it appears that the surfactants of the food simulant solution were adsorbed by the polymer and especially by the laponite particles present in the polymer film or the surface of the polymer film cutting edges. This effect could especially be observed for migration samples using LDPE films with higher laponite loadings. Migration samples prepared from LDPE reference films and LDPE films with only 2% laponite in the polymer caused signals in the AF4/MALLS run exactly like or slightly lower than the surfactant blank which was also stored for 10 days at 60 °C. All migration samples prepared from the LDPE films with 4% laponite in the polymer caused signals significantly lower than the surfactant solution would do without contact to the test film. Migration samples prepared from LDPE films with 6% laponite in the polymer showed no more signal at elution times relevant for laponite particles or the surfactant solution.

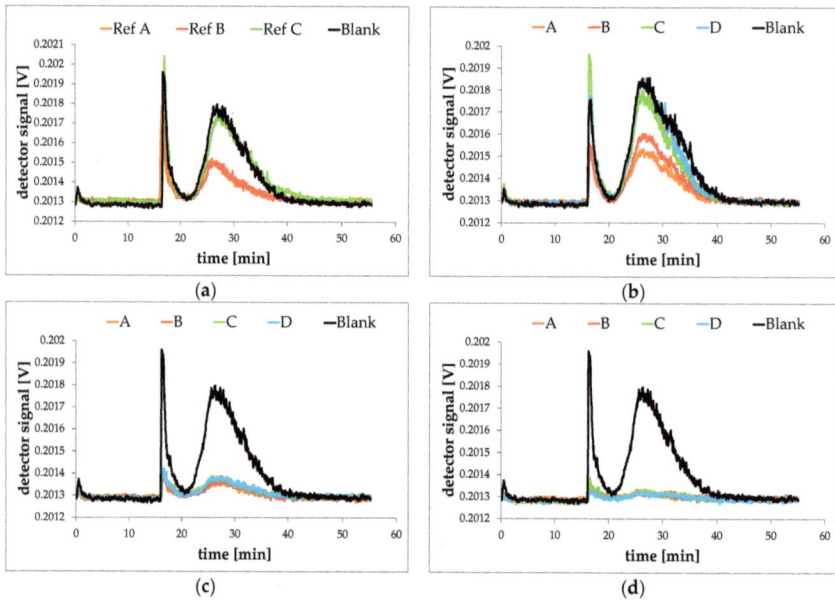

Figure 5. AF4/MALLS fractograms of the migration samples. (**a**) LDPE reference samples without laponite as well as nanocomposites with (**b**) 2% laponite; (**c**) 4% laponite and (**d**) 6% laponite. The black fractograms are the 2000 mg/L Novachem simulant blank that was stored for 10 d at 60 °C.

The preliminary dispersion experiments showed that the Novachem surfactant solution was suitable to disperse laponite particles. Therefore, it is conceivable that the surfactants in the simulant solution were adsorbed by the laponite particles at the surface or the cutting edges of the polymeric films. Beside the AF4/MALLS measurements, the loss of surfactant in the food simulant could be demonstrated by a simple test. Since the surfactant solution is a foam building liquid the presence of the surfactant can be visualized by shaking of the samples (Figure 6). Whilst migration samples using

LDPE blanks and the pure Novachem solution were still foaming, the 2% laponite in LDPE migration samples showed already a lower ability to foam. Samples with 4% and 6% laponite in the polymer showed no foam after they were shaken.

Figure 6. Migration samples after they were shaken. From left to right: 6% Laponite in LDPE, 4% Laponite in LDPE, 2% Laponite in LDPE, LDPE blank and the surfactant blank used as food simulant for all migration samples.

3.4. Validation of Experiments

To validate the migration experiments tests on recovery were performed. Freshly prepared laponite dispersions with 1000 ng/mL laponite were stored for 10 days at 60 °C in 2000 mg/L Novachem solution in the same vials as described. The dispersions in 2000 mg/L Novachem were measured by AF4 and compared to a measurement performed directly after preparation of the laponite dispersion in a Novachem solution (Figure 7). The recovery rates were calculated as the ratio of the detected total peak areas.

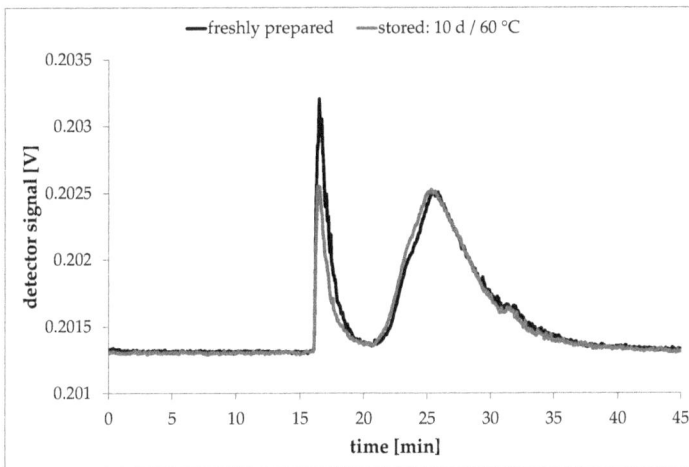

Figure 7. Recovery experiment for 1000 ng/mL Laponite in 2000 mg/L Novachem solution. The black curve is the fractogram of a fresh dispersion; the red curve is the fractogram of the same sample after 10 days at 60 °C (signals of the 68° detector).

The storage of laponite in a Novachem solution showed that the laponite particles have good stability in the surfactant solution. The particles of the stored sample eluted at identical times as the laponite particles from the freshly prepared dispersion and caused similar signal intensities. After storage of the dispersion for 10 days at 60 °C approximately 105% of the original peak area could be recovered (see Table 2). This experiment showed that laponite would have been detectable even after storage for 10 d at 60 °C if migrated into the simulant.

Table 2. Recovery experiment made with a 1000 ng/mL laponite dispersion (2000 μL injections, samples prepared in triplicate).

MALLS Area "Fresh" (mV*min)	MALLS Area "Stored" (mV*min)	Recovery Rate	Direct LOD	LOD Method
182.9/184.3/183.9 183.7 (average)	192.3/197.0/189.7 193.0 (average)	105.1%	500 ng	476 ng

AF4/MALLS measurements can only distinguish between samples of different sizes. For a clear differentiation between polymer, laponite particles and surfactants, the used AF4 method must be able to separate them into separated fractions. In this study, a rapid decline of the separation force (cross flow) was required, where pressure drops caused the surfactant solution to produce a slight peak in the AF4 fractogram at laponite relevant elution times. However, the presence of extracted oligomers from the polymer films or the presence of migrated laponite particles would cause higher signal intensities than dispersant blank. To further validate the method and exclude matrix effects the migration solutions were fortified with a defined amount of laponite. For the validation measurements 10 μL of a 250 μg/mL Laponite dispersion were filled up to the 5 mL mark of a volumetric flask with the respective migration sample. This way, all migration samples were fortified to a laponite content of 500 ng/mL. Injection of a migration solution made from a 6% laponite in LDPE sample was measured. The fractogram showed no signal at the retention time typical for laponite particles. After this run the fortified sample, which was prepared using the identical migration solution than before, was measured. As a result, this AF4 run shows a clear signal at elution times that were typical for the laponite particles (Figure 8).

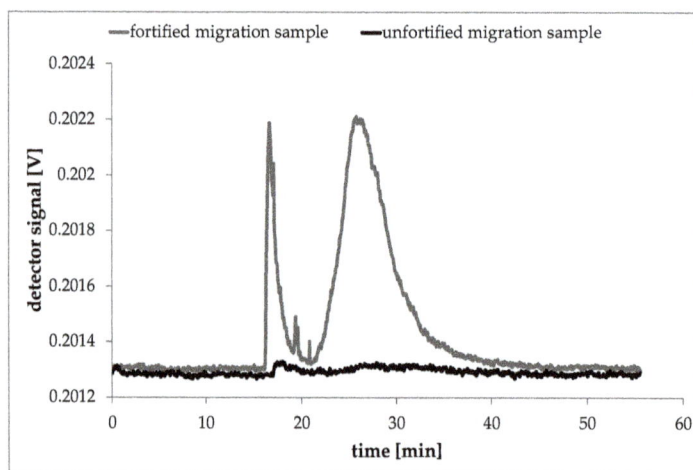

Figure 8. AF4 fractograms from a 6% laponite in LDPE migration solution before and after spiking to 500 ng/mL laponite.

4. Discussion

Due to its size with 1 nm in thickness and 25 nm in diameter only, laponite is one of the smallest clay types (in all three dimensions) and NMs (in one dimension) in general. With migration being based on size-dependent Fickian diffusion [27,34,35] lapoAnite can be considered to be a worst-case nano-additive for FCM made of plastics due to expected higher mobility than most other clay NMs. However, from a migration theory point of view, even laponite platelets are already too large to diffuse within a polymer matrix [27,36].

In this study, no detectable release of laponite particles was found, and even though migration was tested by total immersion of sample pieces where at the cutting edges direct contact of the NM with the used food simulant was possible, this migration test mode should not be applied when NMs would be of small spherical geometry because they may be more easily released from the cut edges into the immersing liquid. Interestingly, it rather appeared that instead of release of laponite particles, the surfactant of the simulant solution was adsorbed. This could happen at best via the cutting edges of the test films where direct contact of laponite with the surrounding simulant matrix was possible. This is further supported by the observed increasing effect with increasing laponite loadings in the polymer. Fortification experiments of migration solutions demonstrated that already low concentrations of laponite would have caused a signal in the AF4/MALLS fractogram. Thus, at the achieved method, detection limit migration of laponite from the nanocomposite could be excluded. From the detection limit of the device, the used amount of simulant and sample area in the migration experiments and the recovery rate under test conditions the overall detection limit of the method was 3.6 µg/dm^2. Assuming a surface-to-volume ratio of 6 dm^2 per kg food, according to the EU cube model, the filling-related detection limit was approximately 22 µg laponite per Kg food.

Other migration studies that investigated clay nanoparticles are summarized in the review of Kuorwel et al. [7]. Though there are studies that reported positive results, the possibility of artefact formation has to be considered [26]. Farhoodi et al. [37] investigated the release of organoclay from a PET-based nanocomposite. In that study, the nanocomposites were cut into several small discs and stored up to 90 d by total immersion in 3% acetic acid. The applied element specific measurements showed the presence of clay-specific elements magnesium, aluminium and silicon. Although the used method based on Inductively Coupled Plasma Optical Emission Spectrometry (ICP-OES) did not allow differentiation between particulate clay and solubilized clay elements (i.e., ionic Mg, Al and Si) the authors concluded that migration of clay was possible. This, of course, cannot be understood as a proof of particle migration. Especially in case of many cutting edges, at which direct contact of clay and the acid is possible and which was the case in that study, great care must be taken to conclude on the actually migrating species. To differentiate between particulate or ionic migrants, suitable particle-specific techniques need to be applied.

In a study by Schmidt et al from 2011 [38] release of organically modified clay from polylactic acid (PLA) matrix was investigated. After storage of the PLA based nanocomposites in 95% ethanol the simulants were digested in acid and analysed for the presence of clay specific elements by Inductively Coupled Plasma Mass Spectrometry (ICP-MS). Likewise, to ICP-OES, normal mode ICP-MS does not allow differentiation between solubilized clay components and particulate clay. However, further particle-specific examinations with a Transmission Electron Microscope (TEM) revealed particulate structures after the acidic digestion solution was evaporated on a TEM sample holder. In such sample preparation steps artefacts might be generated when due to concentration precipitation takes place. Much more importantly, the authors measured the molecular weight (m.w.) of the PLA test samples and found severe changes of the number averaged m.w. of up to 38% during the migration test period. This means that under the aggressive migration test conditions the PLA matrix was largely destroyed by ethanolysis or hydrolysis which lead to a physical degradation of the polymer matrix followed by physical release of clay material [39]. This interpretation would be consistent with the results of a previous study by Schmidt et al. from 2009 [40]. In that study, the authors used a combination of ICP-MS and AF4/MALLS to obtain both element-specific and particle-sensitive information from the migration experiments. In 95% ethanol the authors recorded a signal with AF4/MALLS but could not find clay-specific elements with ICP-MS. The authors concluded therefore that the AF4 signals were caused by released oligomers from the PLA matrix only and not by clay particles.

5. Conclusions

In this study, the potential migration of laponite particles from LDPE-based nanocomposites was investigated. In a first step, the nanomaterial was homogeneously incorporated at different NM loading concentrations into the host LDPE matrix whereby intercalation and partial exfoliation of clay stacks took place. A liquid dispersion with the same size characteristics of laponite was then prepared in an aqueous surfactant solution which was shown to give a stable dispersion of laponite under the used test conditions. This surfactant was therefore used as an appropriate food simulant for migration tests because it can be expected to disperse any migrating laponite particles. Indeed, unlike other chemical polymer additives where olive oil or 95% ethanol are the most severe food simulants, an aqueous surfactant solution can be considered as the most severe simulant in this case because of the good dispersibility and receptivity of laponite NM, once it would be released from the nanocomposite. An analytical method AF4/MALLS was successfully applied for characterization and quantification of laponite particles in the surfactant solution. As a result, no release of laponite, the smallest species of clay, was found in this study. It can therefore be concluded that diffusion-based migration of laponite particles and thus any larger structured nanoclay in general will not occur when the nanomaterial is fully embedded within a polymer matrix. In this study, a potential mechanical release of clays from the surface of the nanocomposite after severe interaction with its environment (e.g., abrasion after chemical, thermal or mechanical impacts) was not investigated. These are additional stress parameters that might appear in practice and should be considered in future research.

Author Contributions: Conceptualization, J.B. and R.F.; Methodology, J.B. and R.F.; Validation, J.B. and R.F.; Investigation, J.B.; Data Curation, J.B.; Writing—Original Draft Preparation, J.B.; Writing—Review & Editing, J.B. and R.F.; Visualization, J.B.; Supervision, R.F.; Project Administration, R.F.

Funding: This research received no external funding.

Acknowledgments: We greatfully acknowledge financial co-support of PlasticsEurope and and FCA (Food Contact Additives), a Sector Group of Cefic, for this study.

Conflicts of Interest: The authors declare no conflict of interest. The study design, the experimental-analytical work, the interpretation of data as well as writing of the manuscript was completely in the responsibility of the authors. The co-funders had no role in this.

References

1. Brigatti, M.F.; Galan, E.; Theng, B.K.G. Structures and mineralogy of clay minerals. In *Handbook of Clayscience*; Bergaya, F., Theng, B.K.G., Lagaly, G., Eds.; Elsevier Ltd.: New York, NY, USA, 2006; pp. 19–86, ISBN 9780080441832.

2. Jatav, S.; Joshi, Y.M. Chemical stability of laponite in aqueous media. *Appl. Clay Sci.* **2014**, *97–98*, 72–77. [CrossRef]

3. Alexandre, M.; Philippe, D. Polymer layered silicate nanocomposites: Preparation, properties and uses of a new class of materials. *Mater. Sci. Eng. R* **2000**, *28*, 1–63. [CrossRef]

4. Azeredo, H. Nanocomposites for food packaging applications. *Food Res. Int.* **2009**, *42*, 1240–1253. [CrossRef]

5. Kaneko, M.L.Q.A.; Romero, R.B.; de Paiva, R.E.F.; Felisberti, M.I.; Gonçalves, M.C.; Yoshida, I.V.P. Improvement of toughness in polypropylene nanocomposite with the addition of organoclay/silicone copolymer masterbatch. *Polym. Compos.* **2013**, *34*, 194–203. [CrossRef]

6. Kiliaris, P.; Papaspyrides, C.D. Polymer/layered silicate (clay) nanocomposites: An overview of flame retardancy. *Prog. Polym. Sci.* **2010**, *35*, 902–958. [CrossRef]

7. Kuorwel, K.K.; Cran, M.J.; Orbell, J.D.; Buddhadasa, S.; Bigger, S.W. Review of mechanical properties, migration, and potential applications in active food packaging systems containing nanoclays and nanosilver. *Compr. Rev. Food Sci. Food Saf.* **2015**, *14*, 411–430. [CrossRef]

8. Paul, D.R.; Robeson, L.M. Polymer nanotechnology: Nanocomposites. *Polymer* **2008**, *49*, 3187–3204. [CrossRef]

9. Pavlidou, S.; Papaspyrides, C.D. A review on polymer–layered silicate nanocomposites. *Prog. Polym. Sci.* **2008**, *33*, 1119–1198. [CrossRef]

10. Pereira de Abreu, D.A.; Paseiro Losadaa, P.; Angulob, I.; Cruza, J.M. Development of new polyolefin films with nanoclays for application in food packaging. *Eur. Polym. J.* **2007**, *43*, 2229–2243. [CrossRef]

11. Sinha Ray, S.; Okamoto, M. Polymer/layered silicate nanocomposites: A review from preparation to processing. *Prog. Polym. Sci.* **2003**, *28*, 1539–1641. [CrossRef]

12. Zhang, W.; Chen, D.; Zhao, Q.; Fang, Y. Effects of different kinds of clay and different vinyl acetate content on the morphology and properties of EVA/clay nanocomposites. *Polymer* **2003**, *44*, 7953–7961. [CrossRef]

13. Bumbudsanpharoke, N.; Ko, S. Nano-food packaging: An overview of market, migration research, and safety regulations. *J. Food Sci.* **2015**, *80*, 910–923. [CrossRef] [PubMed]

14. Agubra, V.A.; Owuor, P.S.; Hosur, M.V. Influence of nanoclay dispersion methods on the mechanical behavior of e-glass/epoxy nanocomposites. *Nanomaterials* **2013**, *3*, 550–563. [CrossRef] [PubMed]

15. Cavallaro, G.; Danilushkina, A.; Evtugyn, V.; Lazzara, G.; Milioto, S.; Parisi, F.; Rozhina, E.; Fakhrullin, R. Halloysite nanotubes: Controlled access and release by smart gates. *Nanomaterials* **2017**, *7*, 199. [CrossRef] [PubMed]

16. Cavallaro, G.; Lazzara, G.; Milioto, S. Sustainable nanocomposites based on halloysite nanotubes and pectin/polyethylene glycol blend. *Polym. Degrad. Stab.* **2013**, *98*, 2529–2536. [CrossRef]

17. Ferrández-Rives, M.; Beltrán-Osuna, Á.A.; Gómez-Tejedor, J.A.; Gómez-Tejedor, J.L. Electrospun PVA/Bentonite Nanocomposites Mats for Drug Delivery. *Materials* **2017**, *10*, 1448.

18. Lazzara, G.; Cavallaroa, G.; Panchalb, A.; Fakhrullinc, R.; Stavitskayad, A.; Vinokurovd, V.; Lvovbd, Y. An assembly of organic-inorganic composites using halloysite clay nanotubes. *Curr. Opin. Colloid Interface Sci.* **2018**, *35*, 42–50. [CrossRef]

19. Makaremi, M.; Pasbakhsh, P.; Cavallaro, G.; Lazzara, G.; Kit Aw, Y.; Mae Lee, S.; Milioto, S. Effect of morphology and size of halloysite nanotubes on functional pectin bionanocomposites for food packaging applications. *ACS Appl. Mater. Interfaces* **2017**, *9*, 17476–17488. [CrossRef] [PubMed]

20. Nikolaidis, A.; Achilias, D. Thermal degradation kinetics and viscoelastic behavior of poly(methyl methacrylate)/organomodified montmorillonite nanocomposites prepared via in situ bulk radical polymerization. *Polymers* **2018**, *10*, 491. [CrossRef]

21. Yoo, J.T.; Lee, S.B.; Lee, C.K.; Hwang, S.W.; Kim, C.R.; Fujigaya, T.; Nakashima, N.; Shim, J.K. Graphene oxide and laponite composite films with high oxygen-barrier properties. *Nanoscale* **2014**, *6*, 10824–10830. [CrossRef] [PubMed]

22. Cavallaro, G.; Lazzara, G.; Milioto, S. Aqueous phase/nanoparticles interface: Hydroxypropyl cellulose adsorption and desorption triggered by temperature and inorganic salts. *Soft Matter* **2012**, *8*, 3627. [CrossRef]

23. Cavallaro, G.; Lazzara, G.; Milioto, S. Dispersions of Nanoclays of Different Shapes into Aqueous and Solid Biopolymeric Matrices. Extended Physicochemical Study. *Langmuir* **2011**, *27*, 1158–1167. [CrossRef] [PubMed]

24. European Union. Commission recommendation of 18 October 2011 on the definition of nanomaterial (2011/696/EU). *Off. J. Eur. Union* **2011**, *275*, 38–40.

25. Duncan, T.V.; Pillai, K. Release of Engineered Nanomaterials from Polymer Nanocomposites: Diffusion, Dissolution, and Desorption. *ACS Appl. Mater. Interfaces* **2015**, *7*, 2–19. [CrossRef] [PubMed]

26. Störmer, A.; Bott, J.; Kemmer, D.; Franz, R. Critical review of the migration potential of nanoparticles in food contact plastics. *Trends Food Sci. Technol.* **2017**, *63*, 39–50. [CrossRef]

27. Franz, R.; Welle, F. Mathematic modelling of migration of nanoparticles from food contact polymers. In *The Use of Nanomaterials in Food Contact Materials—Design, Application, Safety*; Veraart, R., Ed.; Stech Publications Inc.: Lancaster, PA, USA, 2017.

28. European Union. Commission regulation (EU) No. 10/2011 of 14 January 2011 on plastic materials and articles intended to come into contact with food. *Off. J. Eur. Union* **2011**, *L*, 12/1.

29. European Union. Commission Regulation (EU) No. 1282/2011 of 28 November 2011 Amending and Correcting Commission Regulation (EU) No. 10/2011 on plastic materials and articles intended to come into contact with food. *Off. J. Eur. Union* **2011**, *L*, 328/22.

30. European Union. Commission Regulation (EU) No. 1183/2012 of 30 November 2012 amending and correcting Regulation (EU) No. 10/2011 on plastic materials and articles intended to come into contact with food. *Off. J. Eur. Union* **2012**, *L*, 338/11.
31. European Union. Commission Regulation (EU) 2017/752 of 28 April 2017 amending and correcting Regulation (EU) No. 10/2011 on plastic materials and articles intended to come into contact with food. *Off. J. Eur. Union* **2017**, *L*, 113/17.
32. Andersson, M.; Wittgren, B.; Wahlund, K.-G. Accuracy in multiangle Light Scattering Measurements for Molar Mass and Radius Estimations. Model Calculations and Experiments. *Anal. Chem.* **2003**, *75*, 4279–4291. [CrossRef] [PubMed]
33. Podzimek, S. *Light Scattering, Size Exclusion Chromatography and Asymmetric Flow Field Flow Fractionation: Powerful Tools for the Characterization of Polymers, Proteins and Nanoparticles*; John Wiley & Sons, Inc.: Hoboken, NJ, USA, 2011.
34. Kalathi, J.T.; Yamamoto, U.; Schweizer, K.S.; Grest, G.S.; Kumar, S.K. Nanoparticle diffusion in polymer nanocomposites. *Phys. Rev. Lett.* **2014**, *112*, 108301. [CrossRef] [PubMed]
35. Möller, K.; Gevert, T. An FTIR solid-state analysis of the diffusion of hindered phenols in low-density polyethylene (LDPE): The effect of molecular size on the diffusion coefficient. *J. Appl. Polym. Sci.* **1994**, *51*, 895–903. [CrossRef]
36. Bott, J.; Störmer, A.; Franz, R. A model study into the migration potential of nanoparticles from plastics nanocomposites for food contact. *Food Packag. Shelf Life* **2014**, *2*, 73–80. [CrossRef]
37. Farhoodi, M.; Mousavi, S.M.; Sotudeh-Gharebagh, R.; Emam-Djomeh, Z.; Oromiehie, A. Migration of aluminum and silicon from PET/Clay nanocomposite bottles into acidic food simulant. *Packag. Technol. Sci.* **2013**, *27*, 161–168. [CrossRef]
38. Schmidt, B.; Katiyar, V.; Plackett, D.; Larsen, E.H.; Gerds, N.; Bender Koch, C.; Petersen, J.H. Migration of nanosized layered double hydroxide platelets from polylactide nanocomposite films. *Food Addit. Contam.* **2011**, *28*, 956–966. [CrossRef] [PubMed]
39. Glover, R.D.; Miller, J.M.; Hutchison, J.E. Generation of metal nanoparticles from silver and copper objects: Nanoparticle dynamics on surfaces and potential sources of nanoparticles in the environment. *ACS Nano* **2011**, *5*, 8950–8957. [CrossRef] [PubMed]
40. Schmidt, B.; Petersen, J.H.; Bender Koch, C.; Plackett, D.; Johansen, N.R.; Katiyar, V.; Larsen, E.H. Combining asymmetrical flow field-flow fractionation with light-scattering and inductively coupled plasma mass spectrometric detection for characterization of nanoclay used in biopolymer nanocomposites. *Food Addit. Contam.* **2009**, *26*, 1619–1627. [CrossRef] [PubMed]

© 2018 by the authors. Licensee MDPI, Basel, Switzerland. This article is an open access article distributed under the terms and conditions of the Creative Commons Attribution (CC BY) license (http://creativecommons.org/licenses/by/4.0/).

nanomaterials

MDPI

Article

Nanocomposite Zinc Oxide-Chitosan Coatings on Polyethylene Films for Extending Storage Life of Okra (*Abelmoschus esculentus*)

Laila Al-Naamani [1,2], Joydeep Dutta [3] and Sergey Dobretsov [1,4,*]

[1] Department of Marine Science and Fisheries, Sultan Qaboos University, PO Box 34, Al-Khoud, 123 Muscat, Oman; Lnaamani@hotmail.com
[2] Ministry of Municipalities and Water Resources, 112 Muscat, Oman
[3] Functional Materials, Applied Physics Department, SCI School, KTH Royal Institute of Technology, Kista, SE-164 40 Stockholm, Sweden; joydeep@kth.se
[4] Center of Excellence in Marine Biotechnology, Sultan Qaboos University, PO Box 50, Al-Khoud, 123 Muscat, Oman
* Correspondence: sergey@squ.edu.om; Tel.: +968-2441-3552

Received: 30 May 2018; Accepted: 26 June 2018; Published: 29 June 2018

Abstract: Efficiency of nanocomposite zinc oxide-chitosan antimicrobial polyethylene packaging films for the preservation of quality of vegetables was studied using okra *Abelmoschus esculentus*. Low density polyethylene films (LDPE) coated with chitosan-ZnO nanocomposites were used for packaging of okra samples stored at room temperature (25 °C). Compared to the control sample (no coating), the total bacterial concentrations in the case of chitosan and nanocomposite coatings were reduced by 53% and 63%, respectively. The nanocomposite coating showed a 2-fold reduction in total fungal concentrations in comparison to the chitosan treated samples. Results demonstrate the effectiveness of the nanocomposite coatings for the reduction of fungal and bacterial growth in the okra samples after 12 storage days. The nanocomposite coatings did not affect the quality attributes of the okra, such as pH, total soluble solids, moisture content, and weight loss. This work demonstrates that the chitosan-ZnO nanocomposite coatings not only maintains the quality of the packed okra but also retards microbial and fungal growth. Thus, chitosan-ZnO nanocomposite coating can be used as a potential coating material for active food packaging applications.

Keywords: ZnO nanoparticle; nanocomposite coating; chitosan; antimicrobial; active food packaging

1. Introduction

Fruits and vegetables being perishable due to high water content are susceptible to rapid deterioration soon after harvest, requiring them to be properly packaged and stored if not consumed immediately [1]. High perishability of most fruits and vegetables has led investigators to seek new approaches to improve shelf-life. Okra (*Abelmoschus esculentus*, family Malvaceae) is a rich source of vitamin C, calcium, carotene, vitamin B1, folates and contains dietary fibre [2,3] and is a widely grown and consumed vegetable in African and Arabic countries [4]. However, it has a short postharvest life of about 10 days at temperatures from 1 to 10 °C due to high respiration and water loss rates [5,6]. It is quite sensitive to bruising, desiccation, loss of chlorophyll, chilling injury, loss of tenderness (increased toughness), that usually leads to decaying following postharvest handling [5].

Packaging plays a critical role in food safety and quality acting as a barrier that protects the food from the outer environmental conditions, such as contamination by pathogens and spoilage organisms, chemical and physical hazards, thus slowing the process of deterioration [7,8]. Food packaging with antimicrobial properties has received attention due to the ability to arrest or delay microbiological decay of food products [9]. In antimicrobial packaging materials, antimicrobial substances are loaded in the packaging system to reduce the risk of contamination by pathogens [10,11]. One of the most successful approaches of antimicrobial packaging development is the coating of polymer surfaces with antimicrobial substances. As bacterial contamination occurs on the surface, the incorporation of antimicrobial agents as a coating provides high exposure area with minimal dissociation of biocides into the packaged food [12,13]. Recent studies have demonstrated that antimicrobial packaging improves the safety and quality of food products and helps in reducing the amount of preservatives in the food [14]. The preservative action of antimicrobial packaging is based on the release of the agents from the packaging material, which do the preservative action by direct contact with the food or by releasing to the surrounding space [14].

Natural polymers, such as chitosan, starch, clay, and pectin, have been used in food packaging due to their biodegradability, non-toxicity and biological properties [15–17]. Chitosan is non-toxic, film forming, and biodegradable biopolymer with antimicrobial properties, which make it ideal for the development of antimicrobial food packaging [17,18]. It is widely used in food production as a fining agent for clarification and de-acidification of fruit juices and for water purification [19]. Chitosan as a packaging material either alone or with other compounds was proved to improve shelf life while maintaining the quality of meat [20], fish [21] and vegetables [17]. Chitosan can form semipermeable film on fruits and vegetables, introducing host resistance to pathogens [22]. Delay in ripening and shelf life prolongation was observed in fruits and vegetables treated with chitosan [23]. Chitosan is usually blended with other polymers or antimicrobial agents, such as natural extracts and metal oxides to improve its mechanical resistance and antimicrobial properties [24,25]. For example, chitosan was blended with other active substances, such as gallic acid [26], grapefruit seed extract [27], poly-vinyl alcohol [17] and silver nanoparticles [28] to form effective antimicrobial packaging materials.

Metal oxides like titanium dioxide (TiO_2), zinc oxide (ZnO) and magnesium oxide (MgO), have been reported to render antibacterial activity with higher stability in comparison to organic antimicrobial agents [29,30]. In comparison to other metal oxides, ZnO nanoparticles are considered as safe materials for human beings [31]. Zinc is a ubiquitous trace metal and essential for a large number of metalloenzymes in living organisms [7]. Furthermore, ZnO is less toxic than other nanoparticles, such as silver nanoparticles, thus is widely used in the food industry as a supplement for zinc [7,32]. ZnO has been incorporated into the linings of food cans for meat, fish, corn and peas to preserve colour and prevent spoilage [32]. Interest has been arisen on using ZnO nanoparticles as food additives or incorporating them with packaging materials in order to provide antimicrobial properties [7,33,34]. Several studies proposed methods used to incorporate ZnO nanoparticles with low density polyethylene (LDPE), polypropylene (PP) and chitosan [32]. The application of chitosan coating in food was mostly presented as an edible or direct coating. Few reports proposed the use of chitosan films blended with different components for food packaging [17–21]. A recent study by Rahman and co-workers [20] proved the efficiency of chitosan-ZnO nanocomposite films formed into pouches in extending the shelf-life of raw meat. However, to the best of our knowledge, there are no reports on the use of chitosan-ZnO coated packaging materials and their application in food preservation.

In this study, the effect of low density polyethylene (LDPE) packaging films coated with chitosan and chitosan-ZnO nanocomposite on the shelf life and quality of okra (*Abelmoschus esculentus*) was investigated. The specific aims of this study were to: (1) prepare LDPE packaging coated with chitosan and chitosan-ZnO nanocomposites; (2) assess the quality of okra samples stored using active packaging through the evaluation of the microbiological, chemical and physical attributes of the vegetable.

2. Materials and Methods

2.1. Sample Preparation

Fresh okra (*Abelmoschus esculentus*) samples were purchased from a local supermarket (As Seeb, Muscat, Oman). Samples were transported to the marine research laboratory and then they were graded visually for their uniformity in size, shape and brightness of colour. Pieces with average size of 13 cm × 2 cm (length × width) were selected for the experiment (Figure 1). Only vegetables free from insects, defects and visible blemishes were selected and used for the experiments (see below).

Figure 1. Photo of all the Okra samples in different polyethylene packages.

2.2. Preparation and Characterisation of Coatings

LDPE films coated with chitosan and chitosan-ZnO nanocomposites were prepared as previously described by Al-Naamani and co-workers [35]. Briefly, 2% chitosan solution was prepared by dissolving 2 g of chitosan powder (Sigma Aldrich, St. Louis, MO, USA) in 0.5% acetic acid. Then, commercial ZnO nanoparticles (35–45 nm) (Sigma Aldrich, St. Louis, MO, USA) were added to the previously prepared chitosan solution to obtain chitosan-ZnO nanocomposite. Clean LDPE films (5 × 8 × 2 cm) (Cole-Parmer Instrument Co., Cambridgeshire, UK) were treated using plasma instrument (Plasma Technology GmbH, Herrenberg-Gültstein, Germany) (pressure: 0.2 mbar, O_2: 3–4 standard cubic centimetres per minute (SCCM), power: 50%). The atmospheric dielectric barrier discharge (DBD) plasma operated at 22 kHz eliminating the need for impedance matching that is required for inductively coupled (ICP) plasma, radio frequency low pressure (RFP) plasma systems, making it simpler to use for a variety of applications including ashing of organic constituents, cleaning of electron microscopy sample holders and all sample surfaces, etching or structuring of surfaces as well as for the modification of surface properties (hydrophilic/hydrophobic). The system uses 230 V/16 A power consumption including vacuum pump (approximately 800–1200 W) (Pfieffer, Annecy, France) with power that can be chosen from 10% to 100%. Plasma treatment was used to provide a hydrophilic property to the polyethylene (PE) surface in order to result in a better attachment of chitosan to the PE surface. Oxygen or air plasma is known to removes organic contaminants by chemical reaction with highly reactive oxygen radicals and through ablation by energetic oxygen ions, promotes surface oxidation and hydroxylation (OH groups) thus increasing surface wettability. After the plasma treatment, 6 mL of previously prepared chitosan-ZnO nanocomposite solution was sprayed onto the PE surface (10 cm × 15 cm) and allowed to dry at room temperature (26 °C). PE films coated with 2% chitosan were used for comparison, and uncoated PE was used as a control. The coated films were characterized and their antimicrobial activity was reported [35].

Surface morphology of the nano-composite coating was characterized by JEOL JSM-7200 (JEOL Ltd., Akishima, Tokyo, Japan) field emission scanning electron microscope (FESEM) working at 20 kV. The elemental composition of the coatings was determined using Energy Dispersive Spectrometry (EDS) (Oxford Instruments NanoAnalysis & Asylum Research, UK, High Wycombe, UK). The static water contact angles of the nano-composite coatings were measured using a Theta Lite attension tensiometer (Biolin Scientific, Gothenburg, Sweden) using sessile drop technique in order to determine film hydrophobicity. Fourier Transform Infra-red (FTIR) spectroscopy was used to identify the chemical structure of the composite films and the possible interactions between their components. The coated and uncoated LDPE films were analysed using Attenuated Total Reflection (ATR) attachment in Frontier (FTIR) spectrometer (PerkinElmer, Waltham, MA, USA), in a spectral range from 4000 to 500 cm^{-1}.

For determination of Zn ions concentration leached per cm^2 of coated LDPE, pieces of the coated LDPE (5.5 cm × 2.5 cm) were immersed in 20 mL distilled water and kept under agitation (100 rpm) for the duration of the experiment. Water samples were collected and measured using Inductively Coupled Plasma Optical Emission Spectroscopy (ICP-OES) (Varian 710 ES, Santa Clara, CA, USA) for Zn determination. The analysis was done in 6 replicates.

2.3. Experimental Design

Selected okra samples (see sample preparation) were separated into three equal groups for the study. First group was packed in the LDPE films coated with chitosan. The second group was packed in LDPE coated with chitosan-ZnO nanocomposite. The last group was used as a control and the samples were packed in uncoated LDPE films. All okra samples were then stored for 12 days at room temperature (25 °C) for subsequent quality assessment (Figure 1). To determine the overall quality of the samples, analysis of the chemical, physical and microbiological counts of all okra samples (see below) were carried out upon 4 and 8 and 12 days of storage.

2.4. Microbial Analysis

Total bacteria and total fungal count in packed okra samples was determined according to the method reported by Harrigan [36]. Briefly, 10 g of each samples were aseptically homogenized with 90 mL of peptone water for 1 min. Ten-fold serial dilution of homogenate samples were prepared and 0.1 mL of each of these solutions were added onto nutrient agar (Sigma Aldrich, St. Louis, MO, USA) for bacterial analysis or potato dextrose agar (Sigma Aldrich, St. Louis, MO, USA) for fungal analysis. The inoculated agar plates were incubated at 37 °C for 24 h for bacterial counts. The Petri dishes prepared for fungal counts were incubated at 30 °C for 5 days. Samples were prepared in triplicate, and the counts between 30 and 300 CFU/g were only considered. All the results are expressed as $\log_{10}$ CFU/g (CFU—colony forming units).

2.5. Chemical and Physical Properties of Okra Samples

For the determination of pH, 5 g of okra sample after each treatment was homogenized thoroughly with 45 mL of distilled water using a blender (Panasonic Corporation, Tokyo, Japan) whereupon the pH was determined using a pH meter (WTW, Weilheim, Germany). Total soluble solids concentration was measured with a pocket refractometer (ATAGO, Tokyo, Japan). In order to do this, the samples were first homogenized in a blender (Panasonic Corporation, Tokyo, Japan). Then, homogenised samples were placed on the prism glass of the refractometer to take direct readings. Moisture content (wet basis) was calculated from the change in sample weight (initial weight–final weight) determined using a balance (Sartorius, Goettingen, Germany) after drying the samples in a vacuum oven at 70 °C for 24 h. Okra weight loss was calculated by weighing okra samples before and at the last day of storage. The results, represented as means ± standard deviation of measurements, were obtained from 10 randomly chosen samples per treatment. The difference between the initial and the final weight of the samples was considered as a total weight loss. The results are expressed as percentage loss of the initial weight.

2.6. Statistical Analysis

Data were subjected to the analysis of variance (ANOVA). Before the analysis, assumption of normality of the data was verified using the Shapiro-Wilk test [37]. Source of variation were the different treatments of the packaging film and storage time. Significant difference between different measurements were determined by Fisher Least Significant Difference (LSD) post hoc test, at a significant level $p = 0.05$.

3. Results

3.1. Characterisation of Coated PE Films

The FTIR spectra of uncoated LDPE, and LDPE coated with chitosan and chitosan-ZnO nanocomposite are shown in Figure 2. The LDPE spectrum observed in this study was similar to one reported previously [38]. Methylene (CH_2) groups corresponding to the stretching modes at 2920 and 2850 cm^{-1} and deformations at 1464 and 719 cm^{-1} as shown in Figure 2 are well known in PE. After the plasma treatment, new peaks at 1720 cm^{-1} corresponding to C=O stretching vibration and at the region of 3200–3800 cm^{-1} corresponding to hydroxyl group (–OH) vibration can be usually observed. For the chitosan coated PE films, the characteristic peaks of chitosan were observed at 3329 cm^{-1} (N–H and O–H stretching) and at 1649 and 1562 cm^{-1} (amide I and amide II) (Figure 2c). A peak at 1035 cm^{-1} was attributed to the stretching vibration of C–O–C [39], which suggests that chitosan is chemically bonded to polyethylene. In comparison, the spectrum of chitosan-ZnO nanocomposite coating has a slight shift of the bands corresponding to hydroxyl, amino, and amide groups towards lower spectral ranges (Figure 2b). This is attributed to the interaction between chitosan and ZnO nanoparticles.

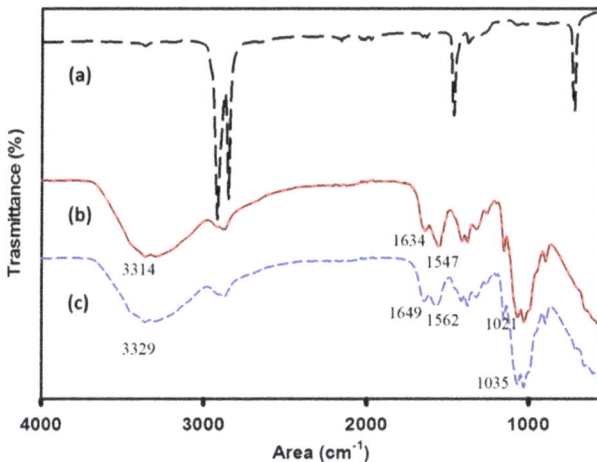

Figure 2. Fourier transform infrared FTIR spectra of low density polyethylene films (LDPE) films coated with chitosan and chitosan-ZnO nanocomposite compared to uncoated LDPE: (**a**) uncoated LDPE; (**b**) LDPE coated with chitosan-ZnO nanocomposite; (**c**) LDPE coated with chitosan.

Scanning Electron micrographs (SEM) confirmed the agglomeration of ZnO nanoparticles to about 500 nm in the nanocomposite coating (Figure 3a). The nanocomposite coating exhibits a hydrophobic surface with a water contact angle of ~95° (Figure 3b). The EDS profile proves that ZnO nanoparticles were successfully incorporated into the chitosan matrix as peaks of zinc were shown in the spectra (Figure 3c). ICP analysis demonstrated that total Zn^{2+} ion concentration that leached out from the nanocomposite coating was 0.00147 ± 0.00008 mg/cm^2 when it was kept under agitation for the duration of the experiment.

(a)

WCA≈ 95°±3.6°

(b)

(c)

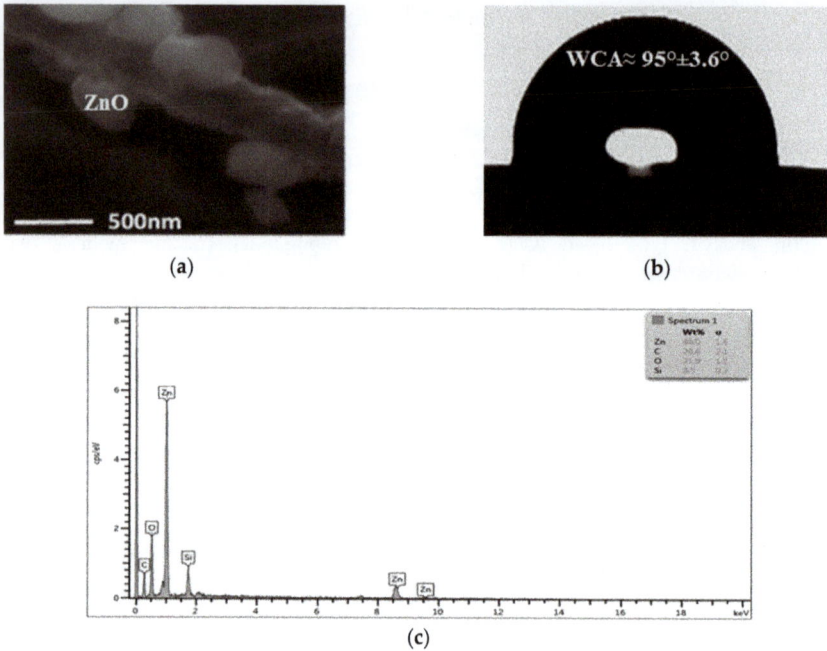

Figure 3. Characterisation of LDPE films coated with chitosan-ZnO nanocomposite. (a) SEM image of the coated LDPE (20,000 ×); (b) Measurement of static contact angle of a water droplet on coated LDPE, the data are means ± standard deviations of five replicates; (c) Energy Dispersive Spectrometry (EDS) spectrum of LDPE surface coated with Chitosan-ZnO nanocomposite. Each peak represents different elements.

3.2. Microbial Analysis of Packed Okra

The bacterial concentration in packed okra samples varied during the whole storage period (Figure 4a). In the first 4 days of storage, there was no significant difference (ANOVA, LSD, $p > 0.05$) in the bacterial concentration between the control (LDPE films) or either of the treated PE films (Figure 4a). After 8 days, the bacterial CFUs increased dramatically in all the samples. Compared to the control sample, the bacterial concentrations in the case of chitosan and nanocomposite coatings were reduced by 53% and 63% respectively, though there was no significant difference (ANOVA, LSD, $p > 0.05$) between the treatments. At the end of experiment, the concentrations of bacteria in all the samples decreased. Similar to day 8, both coatings lead to the reduction of bacterial counts in okra compared to the control (Figure 4a). There was no significant difference (ANOVA, LSD, $p > 0.05$) between the bacterial concentrations in the treated samples.

The fungal concentrations did not differ significantly (ANOVA, LSD, $p > 0.05$) in the first 4 days of storage in all the samples (Figure 4b). Similar to the bacterial counts, fungal concentrations in the samples stored with chitosan and nanocomposite coated films were significantly different (ANOVA, LSD, $p < 0.05$) from the control experiments upon 8 and 12 days of storage. There was no significant difference (ANOVA, LSD, $p > 0.05$) between the fungal concentrations in the chitosan and the nanocomposite samples after 4 and 8 days, but after 12 days fungal counts in samples stored in nanocomposite coated films decreased more than 2-folds (ANOVA, LSD, $p < 0.05$) in comparison to the samples stored in chitosan coated LDPE films. In the control samples the fungal concentrations increased incrementally with storage time (Figure 4b).

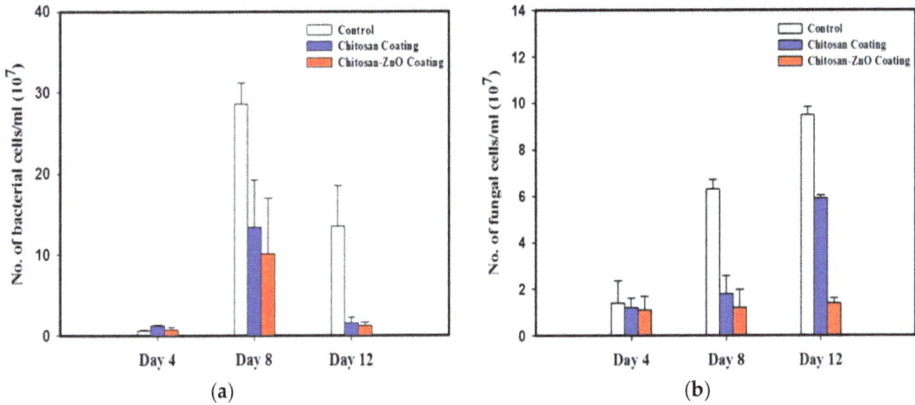

Figure 4. Number of (**a**) bacterial and (**b**) fungal cells (CFU/ml) in okra samples packed in coated and uncoated LDPE films, and incubated for 4, 8 and 12 days.

3.3. Chemical and Physical Properties of Packed Okra

The pH values of okra samples packed with uncoated LDPE, chitosan coated LDPE and LDPE coated with chitosan/ZnO nanocomposite films are presented in Table 1. There was a slight increase in acidity of all of packed okra samples during the experiments. After 12 days of storage, pH values decreased by 1.08%, 2.6%, 1.85% in samples packed in control LDPE films or chitosan and nanocomposite coated films, respectively with no significant difference (ANOVA, LSD, $p > 0.05$).

Table 1. Effect of different coatings on pH of okra samples at each storage periods.

Coating Material	Storage Duration (Days)			
	0	4	8	12
Control (uncoated LDPE)	6.47 ± 0.06 [a,x]	6.44 ± 0.05 [a,x]	6.40 ± 0.07 [a]	6.40 ± 0.01 [a]
Chitosan coating	6.47 ± 0.06 [a,x]	6.43 ± 0.03 [a,x]	6.30 ± 0.06 [b,x]	6.30 ± 0.07 [b,x]
Chitosan/ZnO coating	6.47 ± 0.06 [a,x]	6.34 ± 0.02	6.29 ± 0.07 [b,x]	6.35 ± 0.03 [ab,x]

Note: $\pm$ = standard deviation; Values followed by the same letters in a column (a,b) or in a row (x,y) do not differ significantly.

The amount of total soluble solids in the okra samples increased during the experiment over the period of storage (Table 2). After 4 days of storage, the amount of total soluble solids in the control samples were significantly lower (ANOVA, LSD, $p < 0.05$) than that in the samples stored in either chitosan or nanocomposite coated LDPE films (Table 2). However, all the values increased in the eighth day with no significant difference (ANOVA, LSD, $p > 0.05$) between them. After 12 days, the amount of total soluble solids in the sample with chitosan-ZnO nanocomposite coating was significantly lower (ANOVA, LSD, $p < 0.05$) than the control (Table 2).

Table 2. Effect of different coatings on total soluble solids (brix) in okra samples during different storage periods.

Coating Material	Storage Duration (Days)			
	0	4	8	12
Control (uncoated LDPE)	4.7 ± 0.8 [a,x]	3.6 ± 0.5 [b,x]	5.5 ± 1.0 [a,x]	7.0 ± 0.6 [a]
Chitosan coating	4.7 ± 0.8 [a,x]	4.8 ± 0.9 [a,y]	4.9 ± 0.4 [a,x]	6.0 ± 0.5 [ab]
Chitosan/ZnO coating	4.7 ± 0.8 [a,y]	5.3 ± 0.4 [a,x]	5.3 ± 0.2 [a,y]	5.5 ± 0.1 [b]

Note: ± = standard deviation; Values followed by the same letters in a column (a,b) or in a row (x,y) do not differ significantly.

Moisture content of okra stored at room temperature in the three different packaging materials is shown in Table 3. The moisture content in all samples decreased gradually during the experiment. In the first 4 days of storage, moisture content decreased insignificantly (ANOVA, LSD, $p > 0.05$) by 2.3%, 1.99% and 1.6% for the okra stored in LDPE (control) film or chitosan and nanocomposite coated films, respectively. After 8 days, the control sample lost 10-fold more moisture (ANOVA, LSD, $p < 0.05$) than the samples packed stored in coated packages (Table 3). The control sample lost most of the moisture content within the first 8 days of storage. After 12 days okra samples lost 3.4%, 2.9%, and 2.6% of total moisture from the samples stored in control LDPE films or chitosan and nanocomposite coated films, respectively. The difference between the treatments and the control were not significant.

Table 3. Effect of different coatings on moisture content (%) of okra samples during different storage periods.

Coating Material	Storage Duration (Days)			
	0	4	8	12
Control (uncoated LDPE)	86.3 ± 1.1	84.2 ± 0.4	83.4 ± 1.1	83.4 ± 1.6
Chitosan coating	86.3 ± 1.1	84.6 ± 0.7	84.1 ± 0.9	83.3 ± 0.8
Chitosan/ZnO coating	86.3 ± 1.1	85.2 ± 0.9	84.9 ± 0.6	84.0 ± 1.0

Insignificant variation (ANOVA, LSD, $p > 0.05$) in weight loss was observed among okra samples packed in the two coated films and the control after 12 days of storage. Samples exhibit values of 2.17% ± 0.62, 2.32% ± 1.91 and 2.03% ± 1.24 of weight loss when stored in uncoated films, chitosan or nanocomposite coated LDPE films, respectively.

4. Discussion

The efficiency of any antimicrobial packaging material can be evaluated by several factors. Ideal packaging should provide minimal dissociation of the incorporated antimicrobial agent into the packaged food as well as improve food safety by retarding microbial growth without changing its quality attributes [12–14]. In our experiment, chitosan was incorporated with ZnO nanoparticles and coated onto polyethylene films in order to develop antimicrobial packaging for food storage. This fabricated packaging was characterised and tested for its efficiency in increasing shelf life of okra samples.

The dissociation of Zn^{2+} ions from the packaging coating was determined throughout the experiment period. The concentration of zinc in the coating was 0.08 mg/cm^2. After 12 days, the percentage of Zn^{2+} ions released was calculated to be about 1.8% of the total amount of zinc in the coating. These results showed that zinc release from the fabricated packaging material was still very low even though force was applied to intentionally increase release of Zn^{2+} from the coating. This confirms the stability of ZnO nanoparticles in the chitosan matrix. Chitosan was reported to form a network with ZnO nanoparticles when blended together, which could control the release of Zn^{2+} ions into the environment [40]. Zinc concentration value reported in our study was much lower than the lethal dose concentration estimated for humans on comparison with equivalent studies in animals [41]. As reported by SCF [42], the recommended upper intake level of Zn for human is 25 mg/day.

Antimicrobial activity is a very important factor to evaluate the efficiency of the packaging to be used for food storage. The coatings used in our experiment showed good antimicrobial activity by reducing the amount of bacterial and fungal growth in packed okra samples during the twelve days of storage. In previous studies, the antimicrobial effect of direct chitosan coating of fruits and vegetables has been reported [19,43–45]. Chitosan reduced growth of grey mould [46,47], blue mould [48] and black mould [49] in grapes, strawberries and tomatoes. The growth of foodborne bacteria, such as *E. coli* in tomato [44] and *Salmonella* spp. in whole cantaloupe [45], was reported to get controlled upon the use of chitosan. Antimicrobial effect of plastic bags coated with ZnO nanoparticles was reported previously. Li and co-workers [50] reported 30% reduction in *E. coli* count in cut apple stored in ZnO coated polyvinyl chloride (PVC) bags. Similarly, it was reported by Emamifar and co-workers [51] that the application of low density polyethylene (LDPE) packages blended with ZnO nanoparticles reduced total aerobic bacteria and total yeast and mould in fresh orange juice as well as prolonged its shelf life up to 28 days at 4 °C without any negative effects on sensory quality of the juice. In our study ZnO nanoparticles were mixed with chitosan and bags were coated. Since ZnO nanoparticles and chitosan have antimicrobial properties, we expected a synergistic effect. At the same time, the synergistic effect was observed only on reduction of fungal growth in the okra samples but no significant variation could be found in the inhibition of bacterial growth. Malini and co-workers [52] reported strong antibacterial activity of the chitosan-ZnO nanocomposite membrane with higher inhibition of the Gram negative *K. planticola* than Gram positive *Bacillus subtilis*. Liu and Kim [53] reported a similar synergistic effect of chitosan and ZnO nanoparticles against bacteria *E. coli*, *P. aeruginosa*, *S. aureus*, and *B. subtilis*. A study by Rahman and co-workers [20] revealed that raw beef meat packed into pouches made of chitosan films incorporated with 2% ZnO nanoparticles showed complete inhibition of bacterial growth on the sixth day of storage at 4 C. The ZnO concentration used in this study was 20 times higher than the concentration used in our study (0.1%), which can explain the obtained results. Several mechanisms have been proposed for the strong antimicrobial and antifungal activities of chitosan, most of them owing to the interaction of positive charges of chitosan with the negative charges of microorganisms' cells membrane [54]. It was reported that chitosan can have an effect on the permeabilization of cell membranes of some fungal species depending on their membrane fluidity [19]. Antifungal effect of chitosan can be caused by the biological mechanisms, such as inducing morphological changes, structural alterations of the fungal cells and fruit resistance induction to pathogen attacks [55]. The antimicrobial activity of ZnO nanoparticles is probably related to the photocatalytic generation of reactive oxygen species (ROS) on the surface of the nanoparticles [56–58]. ROS and Zn^{2+} ions are supposed to interact with the anionic components of microbial cell wall causing leakage of these components leading finally to cell death [59]. In the case of ZnO-nanocomposite coatings, synergistic activity was probably due to the enhancement of the positive charges of the amino group of chitosan by ZnO which lead to stronger interaction with negatively charged microbial cell wall [20].

Assessment of quality attributes of packed okra samples was done by measuring pH, total soluble solids, moisture contents and weight loss of samples after each storage time. These measurements give indication of efficiency of the fabricated packaging in preserving quality characteristics of okra samples. The results showed that while acidity of okra samples slightly increased during the twelve days of storage, the pH values did not show any significant changes (ANOVA, LSD, $p > 0.05$). Babarinde and Fabunmi [1] had earlier reported a reduction in pH from 6.7 to 6 of okra samples stored in LDPE films for 9 days. However, there were no other reports, either on okra coated with chitosan or stored in chitosan coated films. Similarly, the effect of storage of okra in chitosan-ZnO nanocomposite coated LDPE films has not been investigated. However, an increase in the pH of grapes honey melon was observed when directly coated with chitosan [19]. In comparison, Hernández-Muñoz and co-workers [60] observed a reduction in pH of strawberries coated with chitosan. These differences in the results about the change in acidity of fruits and vegetables could be attributed to their respective organic contents [19].

Total soluble solids indicate the proportion of dissolved solids, such as sugars, acids, amino acids, ascorbic acids and minerals in fruits and vegetables [61,62]. It is a refractometric index usually measured in Brix units which are equal to per cent of soluble solids. The increase in concentration of the total soluble solids in the sample is directly related to the increase in water loss during the storage period. The minimal effect of the chitosan coating in total soluble solids in our experiment could be due to the use of indirect coating on the package films instead of direct coating on food itself. It was reported that retention of total soluble solids in chitosan coated fruits, such as pears [63], strawberries [60] and banana [64] improved. Direct coating on fruits was reported to reduce respiration level which slows down synthesis and use of metabolites leading to slower carbohydrate hydrolysis into sugars leading to a reduction of the concentration of soluble solids [65,66]. However, this process depends on different other factors such as coating thickness, storage conditions and type of fruit and its ripeness stage [65].

Moisture loss in vegetables occurs due to the post-harvest physiological processes, such as respiration and transpiration [1]. The low water loss in all the samples in our experiment could be arise due to the good barrier properties of LDPE to water vapour loss and the ability to reduce respiration rate of vegetables [67–69]. It was reported previously that initial moisture content of okra is 88% and it dropped to 85% after 9 days storage in LDPE [1]. This rate is higher than moisture reduction rate observed in our experiment, which could be attributed to differences in the barrier properties of LDPE films used.

Packaging films is known to lead to the establishment of high relative humidity inside the package due to the reduction in water diffusion to the atmosphere. Thus, transpiration rate is reduced which in turn decrease weight loss in okra [70]. Babarinde and Fabunmi [1] reported 5.8% weight loss in okra after 9 days storage in LDPE when it was stored at 28 °C. The lower loss in moisture (3.9%) was observed when the samples were stored at lower temperatures (15 °C). Difference between this and our results probably due to differences in okra species and packaging films used in our study. Since there were no reports in okra packed LDPE coated with chitosan or chitosan/ZnO nanocomposite, it was not possible to compare the obtained results directly.

From all the results obtained in this work on properties of packed okra upon prolonged storage, it can be observed that the coatings on LDPE package did not influence the chemical and physical properties of okra, such as pH, total soluble solids and moisture content. As mentioned earlier, this could be attributed to the use of indirect coating of chitosan on the plastic film. Direct coating of chitosan on fruits and vegetable are reported to affect the physical and chemical properties of coated grapes, apple, pear, tomato, sweet pepper amongst others [49,71–74]. Chitosan was reported to reduce oxygen and elevating carbon dioxide levels in coated fruits by providing a semipermeable film around them. This can modify internal atmosphere decreasing the respiration level and metabolic activities of fruits and results in ripening delays [23,75,76]. Manipulation of respiration levels can influence properties such as total soluble solids, moisture, weight loss and pH of fruits and vegetables [23]. For example, because of chitosan semipermeable barrier and its reduction of respiration rate, reduced pH and weight loss of coated apples [74] and reduced weight loss of pears [63,73] were reported.

5. Conclusions

This study showed an improvement in the performance of coated LDPE films with chitosan and with chitosan/ZnO nanocomposites for the preservation of quality of okra samples by maintaining moisture content, total soluble solid and pH as well as preventing bacterial and fungal growth in the stored okra samples. Okra pods harvested with minimum handling were reported to have minimum rotting (3.0%) and good appearance for the first 5 days that could be extended upon cold storage for less than 2 weeks [77]. The obtained results proved the effectiveness of the nanocomposite coating on the reduction of fungal growth in the okra samples for up to 12 storage days. Significant reduction in bacterial growth was observed in the samples stored in treated polyethylene films compared to the control, and the nanocomposite coating performed better for the prevention of fungal growth than chitosan alone. It can be concluded that LDPE coating with chitosan-ZnO nanocomposite is a promising technique in which antimicrobial property is added to the films which could influence its possible applications as active food packaging to prolong shelf-life of packed food.

Author Contributions: Laila Al-Naamani performed experiments, data analysis and was responsible for writing and revising the manuscript. Sergey Dobretsov and Joydeep Dutta were responsible for designing experiments, data analysis, reviewing, revising and editing of the manuscript.

Funding: This research was funded by the Research Council of Oman (TRC, RC/AGR/FISH/16/01) and partially by the Sultan Qaboos University—South Africa grant (CL/SQU-SA/18/01).

Acknowledgments: The authors would like to thank Soad for her help in samples preparation and analysis.

Conflicts of Interest: The authors declare no conflict of interest.

References

1. Babarinde, G.O.; Fabunmi, O.A. Effects of packaging materials and storage temperature on quality of fresh Okra (*Abelmoschus esculentus*) fruit. *Agric. Trop. Subtrop.* **2009**, *42*, 151–156.
2. Hosain, M.M.; Jannat, R.; Islam, M.M.; Sarker, M.K.U. Processing and Preservation of Okra Pickle. *Prog. Agric.* **2010**, *21*, 215–222. [CrossRef]
3. Petropoulos, S.; Fernandes, Â.; Barros, L.; Ferreira, I.C. Chemical composition, nutritional value and antioxidant properties of Mediterranean okra genotypes in relation to harvest stage. *Food Chem.* **2018**, *242*, 466–474. [CrossRef] [PubMed]
4. Akhtar, S.; Khan, A.J.; Singh, A.S.; Briddon, R.W. Identification of a disease complex involving a novel monopartite begomovirus with beta-and alphasatellites associated with okra leaf curl disease in Oman. *Arch. Virol.* **2014**, *159*, 1199–1205. [CrossRef] [PubMed]
5. Locascio, S.J. Comparison of cooling and packaging methods to extend the postharvest life of okra. *Proc. Ha. State Hort. Soc.* **1996**, *109*, 285–288.
6. Huang, S.; Li, T.; Jiang, G.; Xie, W.; Chang, S.; Jiang, Y.; Duan, X. 1-Methylcyclopropene reduces chilling injury of harvested okra (*Hibiscus esculentus* L.) pods. *Sci. Hortic.* **2012**, *141*, 42–46. [CrossRef]
7. Abad, M.A. Development of Silver Based Antimicrobial Films for Coating and Food Packaging Applications. Ph.D. Thesis, University of Valencia, Valencia, Spain, February 2014.
8. Grinstead, D. Antimicrobial food packaging: Breakthroughs and benefits that impact food safety. In Proceedings of the International Association for Food Protection (IAFP) Annual Meeting, St. Louis, MO, USA, 31 July–3 August 2016.
9. Mastromatteo, M.; Conte, A.; Del Nobile, M.A. Advances in controlled release devices for food packaging applications. *Trends Food. Sci. Technol.* **2010**, *21*, 591–598. [CrossRef]
10. Quintavalla, S.; Vicini, L. Antimicrobial food packaging in meat industry. *Meat Sci.* **2002**, *62*, 373–380. [CrossRef]
11. Yildirim, S.; Röcker, B.; Pettersen, M.K.; Nilsen-Nygaard, J.; Ayhan, Z.; Rutkaite, R.; Radusin, T.; Suminska, P.; Marcos, B.; Coma, V. Active Packaging Applications for Food. *Compr. Rev. Food Sci. Food Saf.* **2018**, *17*, 165–199. [CrossRef]
12. Ouattara, B.; Simard, R.E.; Piette, G.; Bégin, A.; Holley, R.A. Inhibition of surface spoilage bacteria in processed meats by application of antimicrobial films prepared with chitosan. *Int. J. Food Microbiol.* **2000**, *62*, 139–148. [CrossRef]
13. Lopez-Rubio, A.; Almenar, E.; Hernandez-Muñoz, P.; Lagarón, J.M.; Catalá, R.; Gavara, R. Overview of active polymer-based packaging technologies for food applications. *Food Rev. Int.* **2004**, *20*, 357–387. [CrossRef]
14. Gherardi, R.; Becerril, R.; Nerin, C.; Bosetti, O. Development of a multilayer antimicrobial packaging material for tomato puree using an innovative technology. *LWT-Food Sci. Technol.* **2016**, *72*, 361–367. [CrossRef]
15. Majeed, K.; Jawaid, M.; Hassan, A.; Bakar, A.A.; Khalil, H.A.; Salema, A.A.; Inuwa, I. Potential materials for food packaging from nanoclay/natural fibres filled hybrid composites. *Mater. Des.* **2013**, *46*, 391–410. [CrossRef]

16. Yoshida, C.M.; Maciel, V.B.V.; Mendonça, M.E.D.; Franco, T.T. Chitosan biobased and intelligent films: Monitoring pH variations. *LWT-Food Sci. Technol.* **2014**, *55*, 83–89. [CrossRef]

17. Pereira, V.A.; de Arruda, I.N.Q.; Stefani, R. Active chitosan/PVA films with anthocyanins from *Brassica oleraceae* (Red Cabbage) as time–temperature indicators for application in intelligent food packaging. *Food Hydrocoll.* **2015**, *43*, 180–188. [CrossRef]

18. Aider, M. Chitosan application for active bio-based films production and potential in the food industry: Review. *LWT-Food Sci. Technol.* **2010**, *43*, 837–842. [CrossRef]

19. Irkin, R.; Guldas, M. Chitosan coating of red table grapes and fresh-cut honey melons to inhibit *Fusarium oxysporum* growth. *J. Food Process. Preserv.* **2014**, *38*, 1948–1956. [CrossRef]

20. Rahman, P.M.; Mujeeb, V.A.; Muraleedharan, K. Flexible chitosan-nano ZnO antimicrobial pouches as a new material for extending the shelf life of raw meat. *Int. J. Biol. Macromol.* **2017**, *97*, 382–391. [CrossRef] [PubMed]

21. Ojagh, S.M.; Rezaei, M.; Razavi, S.H. Improvement of the storage quality of frozen rainbow trout by chitosan coating incorporated with cinnamon oil. *J. Aquat. Food Prod. Technol.* **2014**, *23*, 146–154. [CrossRef]

22. Yu, T.; Li, H.Y.; Zheng, X.D. Synergistic effect of chitosan and *Cryptococcus laurentii* on inhibition of *Penicillium expansum* infections. *Int. J. Food Microbiol.* **2007**, *114*, 261–266. [CrossRef] [PubMed]

23. Romanazzi, G.; Feliziani, E.; Baños, S.B.; Sivakumar, D. Shelf life extension of fresh fruit and vegetables by chitosan treatment. *Crit. Rev. Food Sci. Nutr.* **2017**, *57*, 579–601. [CrossRef] [PubMed]

24. Fernandez-Saiz, P.; Ocio, M.J.; Lagaron, J.M. Antibacterial chitosan-based blends with ethylene-vinyl alcohol copolymer. *Carbohydr. Polym.* **2010**, *80*, 874–884. [CrossRef]

25. Siripatrawan, U.; Vitchayakitti, W. Improving functional properties of chitosan films to be used as active food packaging by incorporation with propolis. *Food Hydrocoll.* **2016**, *61*, 695–702. [CrossRef]

26. Schreiber, S.B.; Bozell, J.J.; Hayes, D.G.; Zivanovic, S. Introduction of primary antioxidant activity to chitosan for application as a multifunctional food packaging material. *Food Hydrocoll.* **2013**, *33*, 207–214. [CrossRef]

27. Tan, Y.M.; Lim, S.H.; Tay, B.Y.; Lee, M.W.; Thian, E.S. Functional chitosan-based grapefruit seed extract composite films for applications in food packaging technology. *Mater. Res. Bull.* **2015**, *69*, 142–146. [CrossRef]

28. Bhoir, S.A.; Chawla, S.P. Silver nanoparticles synthesized using mint extract and their application in chitosan/gelatin composite packaging film. *Int. J. Nanosci.* **2016**, 1650022. [CrossRef]

29. Premanathan, M.; Karthikeyan, K.; Jeyasubramanian, K.; Manivannan, G. Selective toxicity of ZnO nanoparticles toward Gram-positive bacteria and cancer cells by apoptosis through lipid peroxidation. *Nanomedicine* **2011**, *7*, 184–192. [CrossRef] [PubMed]

30. De Azeredo, H.M. Antimicrobial nanostructures in food packaging. *Trends Food Sci. Technol.* **2013**, *30*, 56–69. [CrossRef]

31. Stoimenov, P.K.; Klinger, R.L.; Marchin, G.L.; Klabunde, K.J. Metal oxide nanoparticles as bactericidal agents. *Langmuir* **2002**, *18*, 6679–6686. [CrossRef]

32. Espitia, P.J.P.; Soares, N.D.F.F.; dos Reis Coimbra, J.S.; de Andrade, N.J.; Cruz, R.S.; Medeiros, E.A.A. Zinc oxide nanoparticles: Synthesis, antimicrobial activity and food packaging applications. *Food Bioprocess Technol.* **2012**, *5*, 1447–1464. [CrossRef]

33. Chaudhry, Q.; Scotter, M.; Blackburn, J.; Ross, B.; Boxall, A.; Castle, L.; Aitken, R.; Watkins, R. Applications and implications of nanotechnologies for the food sector. *Food Addit. Contam. Part A* **2008**, *25*, 241–258. [CrossRef] [PubMed]

34. Bradley, E.L.; Castle, L.; Chaudhry, Q. Applications of nanomaterials in food packaging with a consideration of opportunities for developing countries. *Trends Food Sci. Technol.* **2011**, *22*, 604–610. [CrossRef]

35. Al-Naamani, L.; Dobretsov, S.; Dutta, J. Chitosan-zinc oxide nanoparticle composite coating for active food packaging applications. *Innov. Food Sci. Emerg. Technol.* **2016**, *38*, 231–237. [CrossRef]

36. Harrigan, W.F. *Laboratory Methods in Food Microbiology*, 3rd ed.; Gulf Professional Publishing: Houston, TX, USA, 1998.

37. Shapiro, S.S.; Wilk, M.B. An analysis of variance test for normality (complete samples). *Biometrika* **1965**, *52*, 591–611. [CrossRef]

38. Pérez-Gago, M.B.; Rhim, J.W. Edible coating and film materials. In *Innovations in Food Packaging*, 2nd ed.; Academic Press: San Diego, CA, USA, 2014; pp. 325–350.

39. Haldorai, Y.; Shim, J.-J. Chitosan-Zinc Oxide hybrid composite for enhanced dye degradation and antibacterial activity. *Compos. Interfaces* **2013**, *20*, 365–377. [CrossRef]

40. Al-Naamani, L.; Dobretsov, S.; Dutta, J.; Burgess, J.G. Chitosan-ZnO nanocomposite coatings for the prevention of marine biofouling. *Chemosphere* **2017**, *168*, 408–417. [CrossRef] [PubMed]

41. Plum, L.M.; Rink, L.; Haase, H. The essential toxin: Impact of zinc on human health. *Int. J. Environ. Res. Public Health* **2010**, *7*, 1342–1365. [CrossRef] [PubMed]

42. *Opinion of the Scientific Committee on Food (SCF) on the Tolerable Upper Intake Level of Zinc*; European Commission: Brussels, Belgium, 2003. Available online: https://ec.europa.eu/food/sites/food/files/safety/docs/sci-com_scf_out177_en.pdf (accessed on 27 June 2018).

43. Li, Y.C.; Sun, X.J.; Yang, B.I.; Ge, Y.H.; Yi, W.A.N.G. Antifungal activity of chitosan on *Fusarium sulphureum* in relation to dry rot of potato tuber. *Agric. Sci. China* **2009**, *8*, 597–604. [CrossRef]

44. Inatsu, Y.; Kitagawa, T.; Bari, M.L.; Nei, D.; Juneja, V.; Kawamoto, S. Effectiveness of acidified sodium chlorite and other sanitizers to control *Escherichia coli* O157: H7 on tomato surfaces. *Foodborne Pathog. Dis.* **2010**, *7*, 629–635. [CrossRef] [PubMed]

45. Chen, W.; Jin, T.Z.; Gurtler, J.B.; Geveke, D.J.; Fan, X. Inactivation of Salmonella on whole cantaloupe by application of an antimicrobial coating containing chitosan and allyl isothiocyanate. *Int. J. Food Microbiol.* **2012**, *155*, 165–170. [CrossRef] [PubMed]

46. Romanazzi, G.; Nigro, F.; Ippolito, A.; Divenere, D.; Salerno, M. Effects of pre-and postharvest chitosan treatments to control storage grey mould of table grapes. *J. Food Sci.* **2002**, *67*, 1862–1867. [CrossRef]

47. Badawy, M.E.; Rabea, E.I. Potential of the biopolymer chitosan with different molecular weights to control postharvest gray mould of tomato fruit. *Postharvest Biol. Technol.* **2009**, *51*, 110–117. [CrossRef]

48. Liu, J.; Tian, S.; Meng, X.; Xu, Y. Effects of chitosan on control of postharvest diseases and physiological responses of tomato fruit. *Postharvest Biol. Technol.* **2007**, *44*, 300–306. [CrossRef]

49. Reddy, M.B.; Angers, P.; Castaigne, F.; Arul, J. Chitosan effects on blackmold rot and pathogenic factors produced by *Alternaria alternata* in postharvest tomatoes. *J. Am. Soc. Hortic. Sci.* **2000**, *125*, 742–747.

50. Li, W.L.; Li, X.H.; Zhang, P.P.; Xing, Y.G. Development of nano-ZnO coated food packaging film and its inhibitory effect on *Escherichia coli* in vitro and in actual tests. *Adv. Mater. Res.* **2011**, *152*, 489–492. [CrossRef]

51. Emamifar, A.; Kadivar, M.; Shahedi, M.; Soleimanian-Zad, S. Evaluation of nanocomposite packaging containing Ag and ZnO on shelf life of fresh orange juice. *Innov. Food Sci. Emerg. Technol.* **2010**, *11*, 742–748. [CrossRef]

52. Malini, M.; Thirumavalavan, M.; Yang, W.Y.; Lee, J.F.; Annadurai, G. A versatile chitosan/ZnO nanocomposite with enhanced antimicrobial properties. *Int. J. Biol. Macromol.* **2015**, *80*, 121–129. [CrossRef] [PubMed]

53. Liu, Y.; Kim, H.I. Characterization and antibacterial properties of genipin-crosslinked chitosan/poly(ethylene glycol)/ZnO/Ag nanocomposites. *Carbohydr. Polym.* **2012**, *89*, 111–116. [CrossRef] [PubMed]

54. Alisashi, A.; Aïder, M. Applications of chitosan in the seafood industry and aquaculture: A review. *Food Bioprocess Technol.* **2012**, *5*, 817–830. [CrossRef]

55. Yu, T.; Yu, C.; Chen, F.; Sheng, K.; Zhou, T.; Zunun, M.; Abudu, O.; Yang, S.; Zheng, X. Integrated control of blue mould in pear fruit by combined application of chitosan, a biocontrol yeast and calcium chloride. *Postharvest Biol. Technol.* **2012**, *69*, 49–53. [CrossRef]

56. Shi, L.E.; Li, Z.H.; Zheng, W.; Zhao, Y.F.; Jin, Y.F.; Tang, Z.X. Synthesis, antibacterial activity, antibacterial mechanism and food applications of ZnO nanoparticles: A review. *Food Addit. Contam. Part A* **2014**, *31*, 173–186. [CrossRef] [PubMed]

57. Noshirvani, N.; Ghanbarzadeh, B.; Mokarram, R.R.; Hashemi, M.; Coma, V. Preparation and characterization of active emulsified films based on chitosan-carboxymethyl cellulose containing zinc oxide nano particles. *Int. J. Biol. Macromol.* **2017**, *99*, 530–538. [CrossRef] [PubMed]

58. Sathe, P.; Laxman, K.; Myint, M.T.Z.; Dobretsov, S.; Richter, J.; Dutta, J. Bioinspired nanocoatings for biofouling prevention by photocatalytic redox reactions. *Sci. Rep.* **2017**, *7*, 3624. [CrossRef] [PubMed]

59. Zhang, Z.-Y.; Xiong, H.-M. Photoluminescent ZnO nanoparticles and theirbiological applications. *Materials* **2015**, *8*, 3101–3127. [CrossRef]

60. Hernandez-Munoz, P.; Almenar, E.; Del Valle, V.; Velez, D.; Gavara, R. Effect of chitosan coating combined with postharvest calcium treatment on strawberry (*Fragaria ananassa*) quality during refrigerated storage. *Food Chem.* **2008**, *110*, 428–435. [CrossRef] [PubMed]

61. Kader, A.A. Flavor quality of fruits and vegetables. *J. Sci. Food Agric.* **2008**, *88*, 1863–1868. [CrossRef]

62. Beckles, D.M. Factors affecting the postharvest soluble solids and sugar content of tomato (*Solanum lycopersicum* L.) fruit. *Postharvest Biol. Technol.* **2012**, *63*, 129–140. [CrossRef]

63. Lin, L.; Wang, B.; Wang, M.; Cao, J.; Zhang, J.; Wu, Y.; Jiang, W. Effects of a chitosan-based coating with ascorbic acid on post-harvest quality and core browning of 'Yali'pears (*Pyrus bertschneideri* Rehd.). *J. Sci. Food Agric.* **2008**, *88*, 877–884. [CrossRef]

64. Kittur, F.S.; Saroja, N.; Tharanathan, R. Polysaccharide-based composite coating formulations for shelf-life extension of fresh banana and mango. *Eur. Food Res. Technol.* **2001**, *213*, 306–311. [CrossRef]

65. Ali, A.; Muhammad, M.T.M.; Sijam, K.; Siddiqui, Y. Effect of chitosan coatings on the physicochemical characteristics of Eksotika II papaya (*Carica papaya* L.) fruit during cold storage. *Food Chem.* **2011**, *124*, 620–626. [CrossRef]

66. Das, D.K.; Dutta, H.; Mahanta, C.L. Development of a rice starch-based coating with antioxidant and microbe-barrier properties and study of its effect on tomatoes stored at room temperature. *LWT-Food Sci. Technol.* **2013**, *50*, 272–278. [CrossRef]

67. Lee, L.; Arul, J.; Lenck, R.; Castaigne, F. A review on modified atmosphere packaging and preservation of fresh fruits and vegetables. Physiological basis and practical aspects. Part, I. *Packag. Technol. Sci.* **1995**, *9*, 1–17. [CrossRef]

68. Zagory, D. Principle and practice of modified atmosphere packaging of horticultural commodities. In *Principles of Modified Atmosphere and Sous Vide Product Packaging*; Farber, J.M., Dodda, K.L., Eds.; Economic Publishing Co. Inc.: Lancaster, PA, USA, 1995; pp. 175–204.

69. Munteanu, B.S.; Paslaru, E.; Zemljic, L.F.; Sdrobis, A.; Pricope, G.M.; Vasile, C. Chitosan coating applied to polyethylene surface to obtain food packaging materials. *Cellul. Chem. Technol.* **2014**, *48*, 565–575.

70. Rai, D.R.; Balasubramanian, S. Qualitative and textural changes in fresh okra pods (*Hibiscus esculentus* L.) under modified atmosphere packaging in perforated film packages. *Food Sci. Technol. Int.* **2009**, *15*, 131–138. [CrossRef]

71. El Ghaouth, A.; Arul, J.; Ponnampalam, R.; Boulet, M. Use of chitosan coating to reduce water loss and maintain quality of cucumber and bell pepper fruits. *J. Food Process. Preserv.* **1991**, *15*, 359–368. [CrossRef]

72. Meng, X.; Li, B.; Liu, J.; Tian, S. Physiological responses and quality attributes of table grape fruit to chitosan preharvest spray and postharvest coating during storage. *Food Chem.* **2008**, *106*, 501–508. [CrossRef]

73. Zhou, R.; Mo, Y.; Li, Y.; Zhao, Y.; Zhang, G.; Hu, Y. Quality and internal characteristics of Huanghua pears (Pyrus pyrifolia Nakai, cv. Huanghua) treated with different kinds of coatings during storage. *Postharvest Biol. Technol.* **2008**, *49*, 171–179. [CrossRef]

74. Shao, X.F.; Tu, K.; Tu, S.; Tu, J. A combination of heat treatment and chitosan coating delays ripening and reduces decay in "Gala" apple fruit. *J. Food Qual.* **2012**, *35*, 83–92. [CrossRef]

75. Olivas, G.I.; Barbosa-Cánovas, G.V. Edible coatings for fresh-cut fruits. *Crit. Rev. Food Sci. Nutr.* **2005**, *45*, 657–670. [CrossRef] [PubMed]

76. Vargas, M.; Pastor, C.; Chiralt, A.; McClements, D.J.; Gonzalez-Martinez, C. Recent advances in edible coatings for fresh and minimally processed fruits. *Crit. Rev. Food Sci. Nutr.* **2008**, *48*, 496–511. [CrossRef] [PubMed]

77. Dhall, R.K.; Sharma, S.R.; Mahajan, B.V.C. Development of post-harvest protocol of okra for export marketing. *J. Food Sci. Technol.* **2014**, *51*, 1622–1625. [CrossRef] [PubMed]

© 2018 by the authors. Licensee MDPI, Basel, Switzerland. This article is an open access article distributed under the terms and conditions of the Creative Commons Attribution (CC BY) license (http://creativecommons.org/licenses/by/4.0/).

nanomaterials

MDPI

Article

Antimicrobial Membranes of Bio-Based PA 11 and HNTs Filled with Lysozyme Obtained by an Electrospinning Process

Valeria Bugatti, Luigi Vertuccio, Gianluca Viscusi and Giuliana Gorrasi *

Department of Industrial Engineering, University of Salerno, via Giovanni Paolo II, 132, 84084 Fisciano (SA), Italy; vbugatti@unisa.it (V.B.); lvertuccio@unisa.it (L.V.); gviscusi@unisa.it (G.V.)
* Correspondence: ggorrasi@unisa.it

Received: 29 January 2018; Accepted: 27 February 2018; Published: 1 March 2018

Abstract: Bio-based membranes were obtained using Polyamide 11 (PA11) from renewable sources and a nano-hybrid composed of halloysite nanotubes (HNTs) filled with lysozyme (50 wt % of lysozyme), as a natural antimicrobial molecule. Composites were prepared using an electrospinning process, varying the nano-hybrid loading (i.e., 1.0, 2.5, 5.0 wt %). The morphology of the membranes was investigated through SEM analysis and there was found to be a narrow average fiber diameter (0.3–0.5 μm). The mechanical properties were analyzed and correlated to the nano-hybrid content. Controlled release of lysozyme was followed using UV spectrophotometry and the release kinetics were found to be dependent on HNTs–lysozyme loading. The experimental results were analyzed by a modified Gallagher–Corrigan model. The application of the produced membranes, as bio-based pads, for extending the shelf life of chicken slices has been tested and evaluated.

Keywords: food packaging; electrospun fibers; nanoencapsulation; controlled release

1. Introduction

The possibility to extend the shelf life of packaged food is a goal that covers several areas of basic and applied research, such as chemistry, microbiology, and materials science. The critical need in the food packaging field is the application of effective methods to inactivate spoilage and foodborne pathogens on the surface of food products [1]. Antimicrobial agents, directly incorporated into food packaging materials, are able to extend the shelf life of packaged foods, thus sustaining its nutritional and sensory qualities [2]. The antimicrobial agents used in food packaging materials generally include inorganic, organic, and biological active molecules. The antimicrobials classified as natural, efficient, and non-toxic are preferred due to the health and ecological concerns. Nanoscale antimicrobial materials attracted much attention due to their improved antimicrobial activities compared with traditional packaging. There are several technological approaches for preparing nanomaterials, and among these the electrospinning process is one of the most attractive methods due to its continuous fabricating capability and simple operating process [3,4]. The obtained materials possess high surface area to volume ratio making it suitable for various applications, such as filters [5,6], absorbing materials [7], textiles [8], sensors [9] and scaffolds for tissue engineering [10–13]. However, the application of electrospinning in the field of food packaging is still less explored [14–16]. Polyamide-11 (PA 11) is a 100% bio-renewable material. It is a high-performance, semi-crystalline, thermoplastic polymer entirely derived from castor oil. When compared to petroleum-based nylons and other conventional plastics, PA 11 has low net CO_2 emissions and global warming potential. Some of the outstanding properties of PA 11 include high impact and abrasion resistance, low specific gravity, excellent chemical resistance, low water absorption, high thermal stability, and capability to be processed over a wide range of temperatures. PA 11 has also excellent dimensional stability, and maintains physical properties over a wide range of temperatures and environments. The possibility to add small amount of nanofillers into polymer matrices is a valuable strategy to obtain novel materials with new properties and added functionalities. Very recently PA 11 has been mixed with clay-based fillers [17–23], carbon-based nanomaterials [24–29], and inorganic particles [30,31] in order to improve its thermal, mechanical, rheological, and electrical properties. In the last few years a new class of natural clays are attracting great interest as fillers for polymers, the halloysite nanotubes (HNTs). They are green materials, cheap, and available in thousands of tons from natural deposits. HNTs have an average length of about 1000 nm, an internal diameter (lumen) of about 10–15 nm, and external diameter of about 50–80 nm. Their general chemical formula is $Al_2Si_2O_5(OH)_4 \times nH_2O$, with a predominant form of hollow tubes, similar to kaolin structure, but with the alumosilicate sheets rolled into tubes [32–35]. The HNTs external surface is composed of Si–O–Si groups, whereas the internal surface consists of a gibbsite-like array of Al–OH groups. HNTs can be dispersed in polymeric matrices without exfoliation, as required for a good dispersion of layered clays, due to the tubular shape and less abundant –OH groups on the surface. Polymeric materials have been filled with these tubular nano-containers [36–39] for the release of specific active molecules (antimicrobial, drugs, essential oils, flame retardant, self-healing, anticorrosion, etc.) in specific environments [40–46]. Very recently, halloysite nanotubes were also used for enzyme immobilization to study the enzymatic activity [47–50]. The present study aims to report the formulation and preparation of bio-based composite membranes based on halloysite as nanocontainers for lysozyme, as natural antimicrobial agent, and PA 11. Lysozyme is a natural antimicrobial agent categorized as GRAS (Generally Recognized as Safe) by the U.S. Food and Drug Administration (FDA), indicating it can be used in food industry without further approval. The nano-hybrid loading into the electrospun solutions was 1.0%, 2.5%, and 5.0%. The technique used for the membrane preparation was electrospinning. The morphology and structure of the membranes were investigated and correlated to the nano-filler loading and the sustained release of lysozyme molecules. Application of the produced membranes as bio-based pads for extending the shelf life of chicken slices has been tested and evaluated.

2. Experimental

2.1. Materials

PA11 with ρ = 1.026 g·cm^{-3} at T = 25 °C, glass transition temperature T_g = 46 °C and melting temperature T_m = 198 °C (CAS 25035-04-5), halloysite nanoclay powders (CAS 1332-58-7), lysozyme powders (CAS 12650-883), hexafluoroisopropanol (HFiP) (CAS 920-66-1). All materials were supplied from Sigma Aldrich (Milan, Italy) and used as received. The preparation of the nano-hybrid HNTs–lysozyme was carried out accordingly to a previously reported procedure [51]. 3 g of lysozyme were dissolved in 30 mL of water at 50 °C for 20 min. The HNTs (3 g) were then added to the lysozyme solution. Ultrasonic processing was performed for 10 min to make HNTs sufficiently dispersed in the lysozyme solution. Vacuum (0.085 MPa) was applied to remove the air between and within the hollow tubules for 15 min. The solution was then taken out from the vacuum and shaken for 5 min. Vacuum was re-applied for 15 min, to remove the trapped air. The HNTs loaded with lysozyme were dried in an oven for 16 h at 50 °C to reach a constant weight. The procedure above described was repeated twice. The content of lysozyme (wt %) in the HNTs–lysozyme hybrid, using the TGA analysis, was estimated to be around 50 wt %. The lysozyme content much exceed the loading capacity of the halloysite nanotubes. This detected amount is then relative either to the molecules inside the nanotubes, or to the molecules external to the nanotubes, that concur with different modes to the release (see Section 3, discussion on controlled release of lysozyme).

2.2. Electrospinning Procedure

The solutions for electrospinning were prepared using the following ratio between the different components: 0.3 g of PA11 were added to 2.7 g of HFiP (sample pure PA11); 0.297 of PA11 and 0.003 g of HNTs–lysozyme were added to 2.7 g of HFiP (sample PA11/1% HNTs–lysozyme); 0.292 of PA11 and 0.0075 g of HNTs–lysozyme were added to 2.7 g of HFiP (sample PA11/2.5% HNTs–lysozyme); 0.285 of PA11 and 0.015 g of HNTs–lysozyme were added to 2.7 g of HFiP (sample PA11/5% HNTs–lysozyme). The mixed solutions were placed into a 5 mL plastic syringe. An electrode lead of a high voltage power supply (HV Power Supply, Gamma High Voltage Research, Ormond, FL, USA) was connected to the needle tip (internal diameter 0.84 mm) of the syringe. A constant positive DC voltage potential was fixed at 16 kV. A syringe pump (NE-1000 Programmable Single Syringe Pump, New Era Pump Systems Inc., Farmingdale, NY, USA) was used to feed the needle with polymer solution at volumetric flow rate of 1 mL/h. For collecting of the fibrous mats, aluminum plates of 10×10 mm^2 were placed on a grounded aluminum collector, the distance between the collector and the syringe was 20 cm. The solutions were subjected to electrospinning in a closed chamber where the temperature was controlled at 24 °C and the relative humidity at 35%. Table 1 reports the samples and the average fibers diameters.

Table 1. Values of elastic modulus (MPa) for all samples (data plotted in Figure 3).

Filler Loading (wt %)	E (MPa)
0	23 ± 8
1.0	27 ± 16
2.5	58 ± 12
5.0	80 ± 18

2.3. Methods of Analysis

SEM analysis on the electrospun samples was performed with a LEO 1525 microscope. The fiber diameters were detected, in the SEM images, through one-by-one localization, using the software Sigma SCAN (Analyze Images Automatically) considering 500 fibers for every system (Systat Software Inc., San Jose, CA, USA). The use of the software "OriginLab" (Systat Software Inc., San Jose, CA, USA) allowed the evaluation of distribution of diameters for all systems.

Fourier transform infrared (FT-IR) were recorded using a Bruker spectrometer, model Vertex 70 (Bruker Italia, Milano, Italy) (average of 32 scans, at a resolution of 4 cm^{-1}).

The mechanical properties of the samples were evaluated from stress–strain curves obtained using a dynamometric apparatus INSTRON 4301 (ITW Test and Measurement Italia S.r.l., Pianezza, Italy). The experiments were conducted at room temperature with the deformation rate of 5 mm/min. The initial length of the samples was about 10 mm. Elastic moduli were derived from the linear part of the stress–strain curves, giving the sample a deformation of 0.1%. Reported results are the average of data obtained from five samples.

The release kinetics of the lysozyme were performed by ultraviolet spectrometric measurement using a Spectrometer UV-2401 PC Shimadzu (Shimadzu, Kyoto, Japan). The tests were performed using rectangular specimens of 4 cm^2 and the same thickness (150 µm), placed into 25 mL physiological solution and stirred at 100 rpm in an orbital shaker (VDRL MOD. 711+) (ASAL S.R.L., Milan, Italy). The release medium was withdrawn at fixed time intervals and replenished with fresh medium. The considered band was at 265 nm.

Antimicrobial analyses were performed considering 1 g of chicken meat, stored at 4 °C, collected at storage times of 6, 9, and 13 days and added to 9 mL of saline peptone water. The mixture was homogenized for 1 min in a stomacher 400 (Lab Blender, Seward Medical, London, UK) and 1 mL of homogenate subjected to serial dilutions in the same diluent. Aliquots of 0.1 mL of different dilutions were spread onto the culture media selective for *Pseudomonas* (Oxoid, Rodano, Italy). Microorganisms were enumerated with the method based on count of Colony Forming Units (CFU), by using 25–250 CFU plates as range of countable colonies. *Pseudomonas* spp. were selected on *Pseudomonas* (strain PAO1) agar base supplemented with (with selective supplement, CFC) at 30 °C for 72 h (ISO/TS 11059:2009 (IDF/RM 225:2009)) (Oxoid, Rodano, Italy). Chicken breasts were analyzed at 24 h ($t = 0$) postmortem and were cut in small fillets of about $3 \times 1 \times 1$ cm, weight $\cong$ 5–6 g, put on the pads made of unfilled PA11 and PA11 + 5 wt % of HNTs/lysozyme. Samples were stored at 4 °C and examined at intervals of 6, 9, and 13 days. Three samples were analyzed. Reported results are the average of three replicates.

3. Results and Discussion

Figure 1 reports the SEM micrographs of unfilled PA 11 and its composites with various filler loading. It is evident that the electrospinning process, in the experimental condition described in Section 2.2, led in all cases to the formation of randomly oriented, defect-free cylindrical fibers, with a very narrow average fiber diameter (~0.3–0.5 μm).

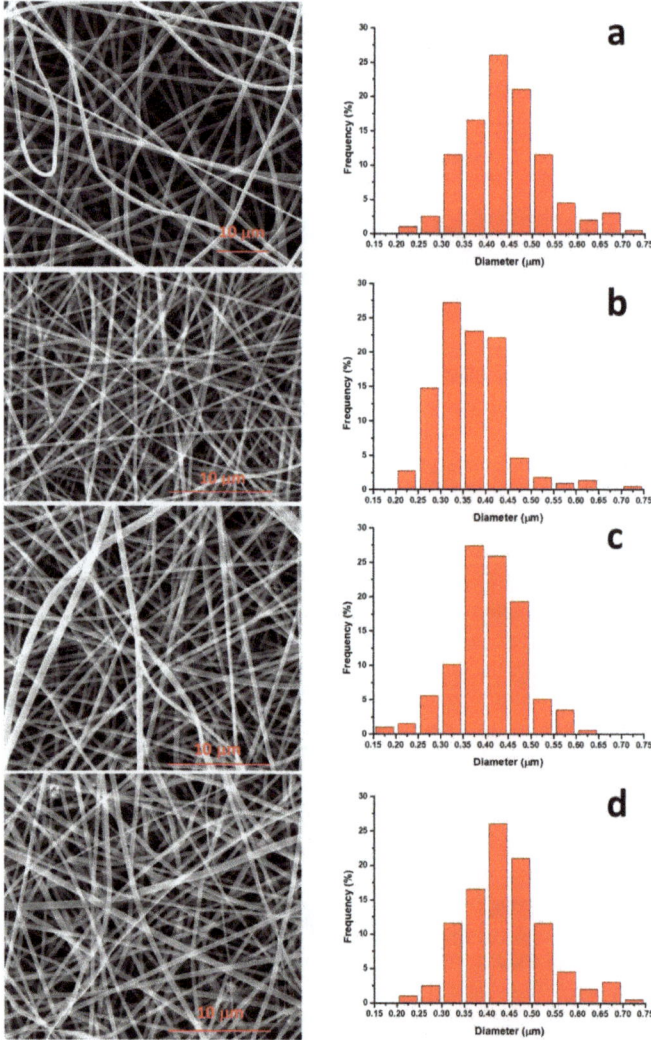

Figure 1. SEM pictures for samples (**a**) PA11, (**b**) PA11/1.0 wt % HNTs–lysozyme, (**c**) PA11/2.5 wt % HNTs–lysozyme, (**d**) PA11/5.0 wt % HNTs–lysozyme.

Figure 2 reports the FTIR spectra of membranes of PA11 and composites in the wavenumber range 2600–3600 cm^{-1}. It can be observed that the band relative to the N–H stretching amide I (3283 cm^{-1}) of the polymer [52] results shifted to higher wavenumber (3297 cm^{-1}) for all the composites. This can be due to hydrogen bonds between the nitrogen of the amidic group of the PA11 and the hydrogen of the carboxyl group of lysozyme molecules external to the HNTs and/or free dispersed into the polymer matrix.

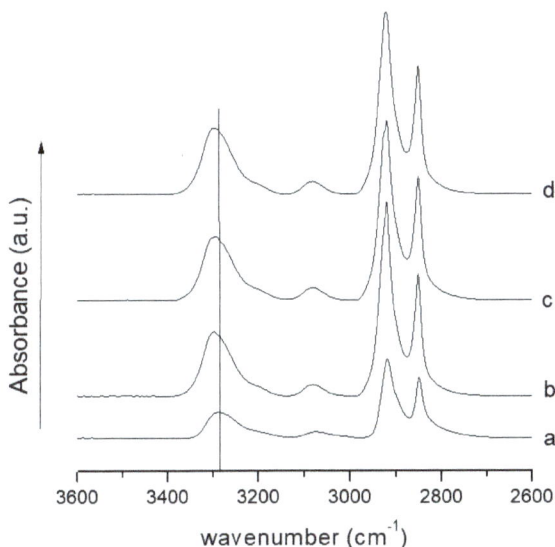

Figure 2. FTIR spectra for samples (a) PA11, (b) PA11/1.0 wt % HNTs–lysozyme, (c) PA11/2.5 wt % HNTs–lysozyme, (d) PA11/5.0 wt % HNTs–lysozyme.

The mechanical properties, in terms of elastic modulus E (MPa), are reported in Figure 3 as a function of filler loading. In the case of composite systems, the enhancement of the mechanical properties of nanocomposites requires a high degree of load transfer between the continuous and dispersed phases. If the interfacial adhesion between the phases is weak, the filler behaves as holes or nanostructured flaws, introducing local stress concentrations, and its benefits on the properties are lost. The filler must be well dispersed, because in the case of poor dispersion, the strength is significantly reduced. The modulus linearly increases with HNTs–lysozyme in all the investigated composition range. The improvement of this parameter can be attributed to the presence of HNTs that act as reinforcing agents for the polymer matrix, well dispersed into the polymer at any composition. Data are reported in Table 1.

Figure 4 reports the release fraction of lysozyme (wt %) versus time (hours) for the considered composite membranes. The release mode for each sample is very complex and a composition of several de-intercalation processes. It is possible to visualize the entire release in three main steps. The initial one is related to the burst and is due to the free lysozyme molecules that are external to the nanofibers and external to the HNTs outside the nanofibers and present on the membranes' surface. The burst's entity increases with the increasing of filler loading in the composites. A second step of release can be attributed to the diffusion of the lysozyme molecules from the bulk of the PA11 nanofibers and external to the HNTs. In the third step, the release mode can be interpreted as desorption and diffusion of lysozyme from the inside of the halloysite nanotubes. It is interesting to note that the total amount of lysozyme released decreases with the filler content, at the same time, and in the investigated range of time (41 days), the release does not reach 100% for the membrane filled with 5 wt % of HNTs–lysozyme (i.e., 2.5% of lysozyme). An empirical equation was adopted, based on a modified Gallagher–Corrigan model [53], that has already been used to describe the lysozyme release from nano-hybrid composite films [46,51]. This model (Equation (1)) presents the combination of two consecutive kinetic mechanisms (dual-model diffusion model). A constant parameter (A) was introduced to take into account the initial burst release. The addition of this parameter simply shifts the model predictions vertically. If no burst occurs, A is zero and the equation is reduced to the original equation.

$$X(t) = A + X_1\left(1 - e^{-C_1 t}\right) + X_2\left(\frac{e^{-C_2(t_{max}-t)}}{\left(1 + e^{-C_2(t_{max}-t)}\right)}\right) \tag{1}$$

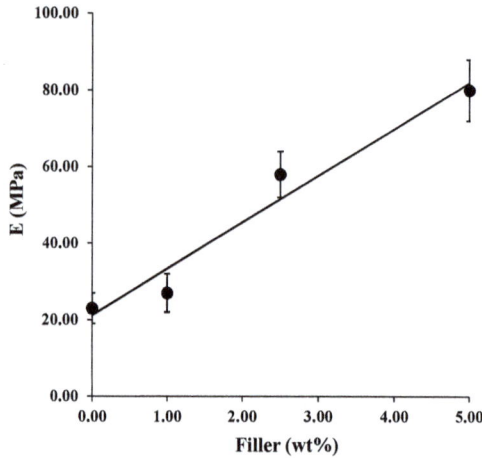

Figure 3. Elastic modulus, E (MPa), as a function of filler content.

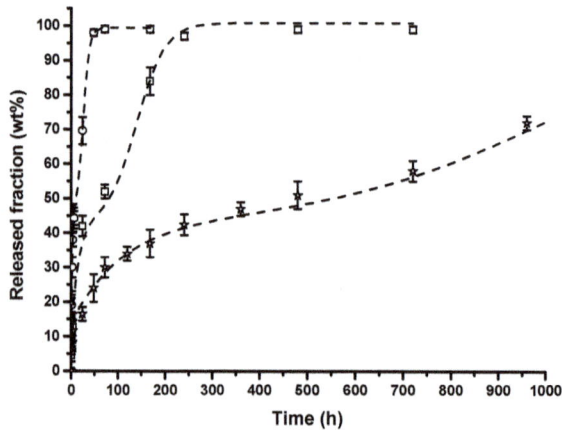

Figure 4. Released fraction (wt %) as a function of time (hour) for samples: (○) PA11/1.0 wt % HNTs–lysozyme, (□) PA11/2.5 wt % HNTs–lysozyme, (☆) PA11/5.0 wt % HNTs–lysozyme.

In Equation (1), $X(t)$ is the fraction of lysozyme released at the time t; X_1 (%) and X_2 (%) are the relative amounts of lysozyme released in the first and second steps of the mechanism; C_1 and C_2 are the kinetic constants of the first and second steps of the release mechanism, t_{max} is time characteristic of the second step mechanism, and A is the burst parameter. Figure 4 also reports the model results for the three analyzed systems as dotted lines. The experimental results for the three considered membranes are well fitted by the model. It is a further confirmation that, apart the initial burst, two distinct steps can occur in the release mechanism. Table 2 reports the model parameters obtained by a nonlinear least squares fitting procedure. Both the rate constant (C_1) and the amount of lysozyme released (X_1) in the initial stage of release slightly decreased with the filler content. It is hypothesized that such behavior is related to the hindrance effect created by the increasing percentage of halloysite nanotubes. The increase in the filler content also generates a delay in the second step time (t_{max}) with a diminution in the resulting rate constant (C_2).

Table 2. Values of the release model parameters evaluated using Equation (1).

Sample	A	X_1 (%)	X_2 (%)	t_{max} (h)	C_1 (h^{-1})	C_2 (h^{-1})	R^2
PA11/1.0 wt % HNTs–lysozyme	6	39	55	25	4.91×10^{-1}	1.46×10^{-1}	0.985
PA11/2.5 wt % HNTs–lysozyme	9	34	58	140	6.57×10^{-2}	3.30×10^{-2}	0.987
PA11/5.0 wt % HNTs–lysozyme	13	30	57	984	9.07×10^{-3}	4.37×10^{-3}	0.985

In order to test the antimicrobial activity against *Pseudomonas*, we used the prepared composites at 5 wt % of HNTs–lysozyme as pads for chicken meat. *Pseudomonas* spp. are specific microbial species which are representative of the microbial dynamics during meat spoilage [54,55]. They are grouped into psychrotrophic microbial groups [54], commonly linked with fresh meat spoilage and with the ability to grow during storage in air, vacuum, and modified atmosphere packaging. Consequently, they are meat colonizers and an important portion of the spoilage microbiota, being occasionally the dominant organisms.

To simulate the meat storage in the field of large food distribution, we put the prepared membranes as pads inside petri dishes of 4 cm diameter. The filled petri dishes were covered with food grade cellophane film. Figure 5 reports the values of CFU/g of *Pseudomonas* for samples evaluated at 6, 9, and 13 days of storage at 4 °C. A pad made of unfilled PA 11 was used as reference. The starting bacterial count is quite low and after six days, had increased by four orders of magnitude for the membrane pad of PA11 and two orders of magnitude for the pad made of composite filled with 5 wt % of HNTs–lysozyme. The bacterial count increases with the time, being always lower for the sample put on the composite material. It is interesting to note that a plateau value is reached for the sample PA11+ 5 wt % HNTs–lysozyme, one order of magnitude lower than the PA11 at 13 days.

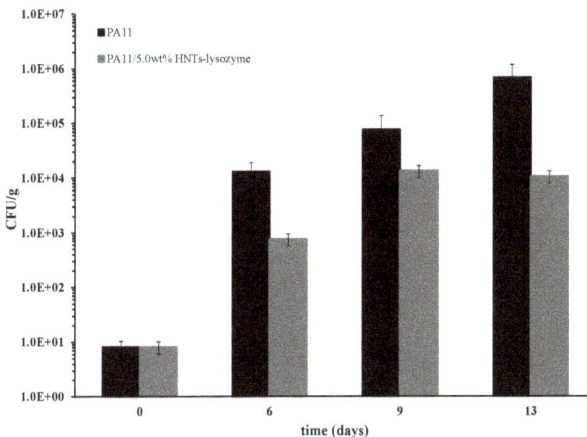

Figure 5. CFU/g of *Pseudomonas* on chicken slices, evaluated at 4 °C, comparing samples PA11 and PA11/5.0 wt % HNTs–lysozyme, as function of storage time.

4. Concluding Remarks

In this paper, the preparation of bio-based membranes composed of Polyamide 11 (PA11) from renewable sources and lysozyme encapsulated into halloysite nanotubes (HNTs) is reported. The lysozyme content of the HNTs–lysozyme hybrid was ≅50 wt %. The hybrid loading into the PA11 was 1.0, 2.5, 5.0 wt %.

- SEM analysis revealed that, with the used processing conditions, both PA11 membrane and the composites show a narrow average fiber diameter (0.3–0.5 μm).
- The FTIR analysis revealed a shift of the N–H stretching of amide I, indicative of a good interaction between the PA11 and lysozyme molecules.
- The mechanical properties, in terms of elastic modulus, increase with filler content for the reinforcing effect of the HNTs.
- The release kinetics of composites' membranes were found to be dependent on the nano-hybrid loading and were well fitted with a modified Gallagher–Corrigan model. It was demonstrated that varying the filler loading it is possible to tune the lysozyme release for desired applications.
- The membranes were used as antimicrobial pads for chicken meat storage. The membrane filled with 5.0 wt % of HNTs–lysozyme was tested against *Pseudomonas* growth for up to 13 days and compared with the unfilled PA11. A reduction in bacterial growth was found for the membrane filled with the antimicrobial compound.

Acknowledgments: This work was supported by the project "High Performing Advanced Material Platform for Active and Intelligent Food Packaging: Cronogard™" (H2020-SMEINST-2-2016-2017). Grant agreement No. 783696. Authors wish to thank Sabrina Carola Carroccio for many fruitful discussions.

Author Contributions: Giuliana Gorrasi conceived the paper and designed the experiments. Gianluca Viscusi, Luigi Vertuccio and Valeria Bugatti performed the experiments. Giuliana Gorrasi, Valeria Bugatti and Luigi Vertuccio analyzed the data. Giuliana Gorrasi wrote this paper.

Conflicts of Interest: The authors declare no conflict of interest.

References

1. Barros-Velazquez, J. *Antimicrobial Food Packaging*, 1st ed.; Elsevier Academic Press: Amsterdam, The Netherlands, 2016.
2. Han, J.H. Antimicrobial packaging systems. In *Plastic Films in Food Packaging*; Ebnesajjad, S., Ed.; Elsevier William Andrew: Waltham, MA, USA, 2013; pp. 151–180.
3. Wanga, G.; Yua, D.; Kelkar, A.D.; Zhang, L. Electrospun nanofiber: Emerging reinforcing filler in polymer matrix composite materials. *Prog. Polym. Sci.* **2017**, *75*, 73–107. [CrossRef]
4. Lim, C.T. Nanofiber technology: Current status and emerging developments. *Prog. Polym. Sci.* **2017**, *70*, 1–17.
5. Cai, J.; Lei, M.; Zhang, Q.; He, J.; Chen, T.; Liu, S.; Fu, S.; Li, T.; Liu, G.; Fei, P. Electrospun composite nanofiber mats of Cellulose@Organically modified montmorillonite for heavy metal ion removal: Design, characterization, evaluation of absorption performance. *Compos. Part A* **2017**, *92*, 10–16. [CrossRef]
6. Essalhi, M.; Khayet, M. Self-sustained webs of polyvinylidene fluoride electrospun nanofibers at different electrospinning times: 1. Desalination by direct contact membrane distillation. *J. Membr. Sci.* **2013**, *433*, 167–179. [CrossRef]
7. Azarniya, A.; Eslahi, N.; Mahmoudi, N.; Simchi, A. Effect of graphene oxide nanosheets on the physico-mechanical properties of chitosan/bacterial cellulose nanofibrous composites. *Compos. Part A* **2016**, *85*, 113–122. [CrossRef]
8. Dhineshbabu, N.R.; Karunakaran, G.; Suriyaprabha, R.; Manivasakan, P.; Rajen-dran, V. Electrospun MgO/Nylon 6 Hybrid Nanofibers for Protective Clothing. *Nano-Micro Lett.* **2004**, *6*, 46–54. [CrossRef]
9. Kadir, R.A.; Li, Z.; Sadek, A.Z.; Rani, R.A.; Zoolfakar, A.S.; Field, M.R.; Ou, J.Z.; Chrimes, A.F.; Kalantar-Zadeh, K. Electrospun granular hollow SnO_2 nanofibers hydrogen gas sensors operating at low temperatures. *J. Phys. Chem. C* **2014**, *118*, 3129–3139. [CrossRef]

10. Lai, K.; Jiang, W.; Tang, J.Z.; Wu, Y.; He, B.; Wang, G.; Gu, Z. Superparamagnetic nano-composite scaffolds for promoting bone cell proliferation and defect reparation without a magnetic field. *RSC Adv.* **2012**, *2*, 13007–13017. [CrossRef]

11. He, L.; Shi, Y.; Han, Q.; Zuo, Q.; Ramakrishna, S.; Zhou, L. Surface modification of electrospun nanofibrous scaffolds via polysaccharide–protein assembly multilayer for neurite outgrowth. *J. Mater. Chem.* **2012**, *26*, 13187–13196. [CrossRef]

12. Yu, D.; Chian, W.; Wang, X.; Li, X.; Li, Y.; Liao, Y. Linear drug release membrane prepared by a modified coaxial electrospinning process. *J. Membr. Sci.* **2013**, *428*, 150–156. [CrossRef]

13. Gao, J.; Zhu, J.; Luo, J.; Xiong, J. Investigation of microporous composite scaffolds fabricated by embedding sacrificial polyethylene glycol microspheres in nanofibrous membrane. *Compos. Part A* **2016**, *91*, 20–29. [CrossRef]

14. Anu Bhushani, J.; Anandharamakrishnan, C. Electrospinning and electrospraying techniques: Potential food based applications. *Trends Food Sci. Technol.* **2014**, *38*, 21–33. [CrossRef]

15. Dìez-Pascual, A.M.; Dìez-Vicente, A.L. Antimicrobial and sustainable food packaging based on poly(butylene adipate-*co*-terephthalate) and electrospun chitosan nanofibers. *RSC Adv.* **2015**, *5*, 93095–93107. [CrossRef]

16. Wen, P.; Zhu, D.; Wu, H.; Zong, M.; Jing, Y.; Han, S. Encapsulation of cinnamon essential oil in electrospun nanofibrous film for active food packaging. *Food Control* **2016**, *59*, 366–376. [CrossRef]

17. Kolesov, I.; Androsch, R.; Mileva, D.; Lebek, W.; Benhamida, A.; Kaci, M.; Focke, W. Crystallization of a polyamide 11/organo-modified montmorillonite nanocomposite at rapid cooling. *Colloid Polym. Sci.* **2013**, *291*, 2541–2549. [CrossRef]

18. Filippone, G.; Carroccio, S.C.; Mendichi, R.; Gioiella, L.; Dintcheva, N.T.; Gambarotti, C. Time-resolved rheology as a tool to monitor the progress of polymer degradation in the melt state—Part I: Thermal and thermo-oxidative degradation of polyamide 11. *Polymer* **2015**, *72*, 134–141. [CrossRef]

19. Filippone, G.; Carroccio, S.C.; Curcuruto, G.; Passaglia, E.; Gambarotti, C.; Dintcheva, N.T. Time-resolved rheology as a tool to monitor the progress of polymer degradation in the melt state—Part II: Thermal and thermo-oxidative degradation of polyamide 11/organo-clay nanocomposites. *Polymer* **2015**, *73*, 102–110. [CrossRef]

20. Kolesov, I.; Androsch, R.; Mileva, D.; Lebek, W.; Benhamida, A.; Kaci, M.; Jariyavidyanont, K.; Focke, W.; Androsch, R. Crystallization kinetics of polyamide 11 in the presence of sepiolite and montmorillonite nanofillers. *Colloid Polym. Sci.* **2016**, *294*, 1143–1151.

21. Risite, H.; El Mabrouk, K.; Bousmina, M.; Fassi-Fehri, O. Role of polyamide 11 interaction with clay and modifier on thermal, rheological and mechanical properties in polymer clay nanocomposites. *J. Nanosci. Nanotechnol.* **2016**, *16*, 7584–7593. [CrossRef]

22. Dintcheva, N.T.; Al-Malaika, S.A.; Morici, E. Novel organo-modifier for thermally-stable polymer-layered silicate nanocomposites. *Polym. Degrad. Stab.* **2015**, *122*, 88–101. [CrossRef]

23. Prashantha, K.; Lacrampe, M.; Krawczak, P. Highly Dispersed Polyamide-11/Halloysite Nanocomposites: Thermal, Rheological, Optical, Dielectric, and Mechanical Properties. *J. Appl. Polym. Sci.* **2013**, *130*, 313–321. [CrossRef]

24. Carponcin, D.; Dantras, E.; Aridon, G.; Levallois, F.; Cadiergues, L.; Lacabanne, C. Evolution of dispersion of carbon nanotubes in Polyamide 11 matrix composites as determined by DC conductivity. *Compos. Sci. Technol.* **2012**, *72*, 515–520. [CrossRef]

25. Jin, J.; Rafiq, R.; Gill, Y.Q.; Song, M. Preparation and characterization of high performance of graphene/nylon nanocomposites. *Eur. Polym. J.* **2013**, *49*, 2617–2626. [CrossRef]

26. Carponcin, D.; Dantras, E.; Dandurand, J.; Aridon, G.; Levallois, F.; Cadiergues, L.; Lacabanne, C. Discontinuity of physical properties of carbon nanotube/polymer composites at the percolation threshold. *J. Non-Cryst. Solids* **2014**, *392*, 19–25. [CrossRef]

27. Rashmi, B.J.; Prashantha, K.; Lacrampe, M.-F.; Krawczak, P. Scalable Production of Multifunctional Bio-Based Polyamide 11/Graphene Nanocomposites by Melt Extrusion Processes via Masterbatch Approach. *Adv. Polym. Technol.* **2016**. [CrossRef]

28. David, C.; Capsal, J.; Laffont, L.; Dantras, E.; Lacabanne, C. Piezoelectric properties of polyamide 11/NaNbO$_3$ nanowire composites. *J. Phys. D Appl. Phys.* **2012**, *45*, 415305. [CrossRef]

29. Naffakh, M.; Shuttleworth, P.S.; Ellis, G. Bio-based polymer nanocomposites based on nylon 11 and WS2 inorganic nanotubes. *RSC Adv.* **2015**, *5*, 17879–17887. [CrossRef]

30. Yuan, P.; Tan, D.; Annabi-Bergaya, F. Properties and applications of halloysite nanotubes: Recent research advances and future prospects. *Appl. Clay Sci.* **2015**, *112*, 75–93. [CrossRef]

31. Lvov, Y.; Wang, W.; Zhang, L.; Fakhrullin, R. Halloysite Clay Nanotubes for Loading and Sustained Release of Functional Compounds. *Adv. Mater.* **2016**, *28*, 1227–1250. [CrossRef] [PubMed]

32. Lvov, Y.; Abdullayev, E. Green and functional polymer-clay nanotube composites with sustained release of chemical agents. *Prog. Polym. Sci.* **2013**, *38*, 1690–1719. [CrossRef]

33. Hillier, S.; Brydson, R.; Delbos, E.; Fraser, T.; Gray, N.; Pendloeski, H.; Phillips, I.; Robertson, J.; Wilson, I. Correlations among the mineralogical and physical properties of halloysite nanotubes (HNTs). *Clay Miner.* **2016**, *51*, 325–350. [CrossRef]

34. Lvov, Y.; Shchukin, D.; Möhwald, H.; Price, R. Clay Nanotubes for Controlled Release of Protective Agents–Perspectives. *ACS Nano* **2008**, *2*, 814–820. [CrossRef] [PubMed]

35. Arcudi, F.; Cavallaro, G.; Lazzara, G.; Massaro, M.; Milioto, S.; Noto, R.; Riela, S. Selective Functionalization of Halloysite Cavity by Click Reaction: Structured Filler for Enhancing Mechanical Properties of Bionanocomposite Films. *J. Phys. Chem. C* **2014**, *118*, 15095–15101. [CrossRef]

36. Gorrasi, G.; Pantani, R.; Murariu, M.; Dubois, P. PLA/Halloysite Nanocomposite Films: Water Vapor Barrier Properties and Specific Key Characteristics. *Macromol. Mater. Eng.* **2014**, *299*, 104–115. [CrossRef]

37. Liu, M.; Jia, Z.; Jia, D.; Zhou, C. Recent advance in research on halloysite-nanotubes polymer nanocomposites. *Prog. Polym. Sci.* **2014**, *39*, 1498–1525. [CrossRef]

38. Abdullayev, E.; Lvov, Y. Clay Nanotubes for Corrosion Inhibitor Encapsulation: Release Control with End Stoppers. *J. Mater. Chem.* **2010**, *20*, 6681–6687. [CrossRef]

39. Abdullayev, E.; Lvov, Y. Clay Nanotubes for Controlled Release of Protective Agents—A Review. *J. Nanosci. Nanotechnol.* **2011**, *11*, 10007–10026. [CrossRef] [PubMed]

40. Scarfato, P.; Avallone, E.; Incarnato, L.; Di Maio, L. Development and evaluation of halloysite nanotube-based carrier for biocide activity in construction materials protection. *Appl. Clay Sci.* **2016**, *132*, 336–342. [CrossRef]

41. Abdullayev, E.; Shchukin, D.; Lvov, Y. Halloysite Clay Nanotubes as a Reservoir for Corrosion Inhibitors and Template for Layer-by-Layer Encapsulation. *Mater. Sci. Eng.* **2008**, *99*, 331–332.

42. Gorrasi, G. Dispersion of halloysite loaded with natural antimicrobials into pectins: Characterization and controlled release analysis. *Carbohydr. Polym.* **2015**, *127*, 47–53. [CrossRef] [PubMed]

43. Gorrasi, G.; Vertuccio, L. Evaluation of zein/halloysite nano-containers as reservoirs of active molecules for packaging applications: Preparation and analysis of physical properties. *J. Cereal Sci.* **2016**, *70*, 66–71. [CrossRef]

44. Gorrasi, G.; Attanasio, G.; Izzo, L.; Sorrentino, A. Controlled release mechanisms of sodium benzoate from a biodegradable polymer and halloysite nanotube composite. *Polym. Int.* **2017**, *66*, 690–698. [CrossRef]

45. Tully, J.; Yendluri, R.; Lvov, Y. Halloysite Clay Nanotubes for Enzyme Immobilization. *Biomacromolecules* **2016**, *17*, 615–621. [CrossRef] [PubMed]

46. Zhai, R.; Zhang, B.; Liu, L.; Xie, Y.; Zhang, H.; Liu, J. Immobilization of enzyme biocatalyst on natural halloysite nanotubes. *Catal. Commun.* **2010**, *12*, 259–263. [CrossRef]

47. Bugatti, V.; Viscusi, G.; Naddeo, C.; Gorrasi, G. Nanocomposites Based on PCL and Halloysite Nanotubes Filled with Lysozyme: Effect of Draw Ratio on the Physical Properties and Release Analysis. *Nanomaterials* **2017**, *7*, 213. [CrossRef] [PubMed]

48. Sun, J.; Yendluri, R.; Liu, K.; Guo, Y.; Lvov, Y.; Yan, X. Enzyme-immobilized clay nanotube-chitosan membranes with sustainable biocatalytic activities. *Phys. Chem.* **2017**, *19*, 562–567. [CrossRef] [PubMed]

49. Bugatti, V.; Sorrentino, A.; Gorrasi, G. Encapsulation of Lysozyme into halloysite nanotubes and dispersion in PLA: Structural and physical properties and controlled release analysis. *Eur. Polym. J.* **2017**, *93*, 495–506. [CrossRef]

50. Gallagher, K.M.; Corrigan, O.I. Mechanistic aspects of the release of levamisole hydrochloride from biodegradable polymers. *J. Control. Release* **2000**, *69*, 261–272. [CrossRef]

51. Qian, M.; Sun, Y.; Xu, X.; Liu, L.; Song, P.; Yu, Y.; Wangd, H.; Qian, J. 2D-Alumina platelets enhance mechanical and abrasion properties of PA612 via interfacial hydrogen-bond interactions. *Chem. Eng. J.* **2017**, *308*, 760–771. [CrossRef]

52. Nychas, G.J.E.; Skandamis, P.N.; Tassou, C.C.; Koutsoumanis, K.P. Meat spoilage during distribution. *Meat Sci.* **2008**, *78*, 77–89. [CrossRef] [PubMed]

53. Gill, C.O.; Newton, K.G. The ecology of bacterial spoilage of fresh meat at chill temperatures. *Meat Sci.* **1978**, *14*, 43–60. [CrossRef]
54. Jay, J.M.; Loessner, M.J.; Golden, D.A. *Modern Food Microbiology*; Springer: Berlin, Germany, 2005; Chapter 4; pp. 63–91.
55. Gorrasi, G.; Milone, C.; Piperopoulos, E.; Lanza, M.; Sorrentino, A. Hybrid clay mineral-carbon nanotube-PLA nanocomposite films. Preparation and photodegradation effect on their mechanical, thermal and electrical properties. *Appl. Clay Sci.* **2013**, *71*, 49–54. [CrossRef]

© 2018 by the authors. Licensee MDPI, Basel, Switzerland. This article is an open access article distributed under the terms and conditions of the Creative Commons Attribution (CC BY) license (http://creativecommons.org/licenses/by/4.0/).

MDPI

St. Alban-Anlage 66

4052 Basel

Switzerland

Tel. +41 61 683 77 34

Fax +41 61 302 89 18

www.mdpi.com

Nanomaterials Editorial Office

E-mail: nanomaterials@mdpi.com

www.mdpi.com/journal/nanomaterials

www.ingramcontent.com/pod-product-compliance
Lightning Source LLC
Chambersburg PA
CBHW051914210326
41597CB00033B/6142